U0318525

# URBAN
# LANDSCAPE PLANNING

# 城市景观规划

深圳市艺力文化发展有限公司 编
君誉文化 策划

华南理工大学出版社
SOUTH CHINA UNIVERSITY OF TECHNOLOGY PRESS
· 广州 ·

图书在版编目（CIP）数据

城市景观规划 = Urban landscape planning ：英汉对照 / 深圳市艺力文化发展有限公司编. — 广州 ：华南理工大学出版社，2017.9
ISBN 978-7-5623-4527-5

Ⅰ.①城… Ⅱ.①深… Ⅲ.①城市景观－景观规划－英、汉 Ⅳ.①TU-856

中国版本图书馆CIP数据核字（2015）第008278号

Urban Landscape Planning
城市景观规划
深圳市艺力文化发展有限公司　编　　君誉文化　策划

---

出 版 人：卢家明
出版发行：华南理工大学出版社
（广州五山华南理工大学17号楼，邮编510640）
http://www.scutpress.com.cn　E-mail: scutc13@scut.edu.cn
营销部电话：020-87113487　87111048（传真）
策划编辑：赖淑华
责任编辑：陈　昊　黄丽谊
印 刷 者：上海锦良印刷厂有限公司
开　　本：595mm × 1020mm　1/16　印张：17
版　　次：2017年9月第1版　2017年9月第1次印刷
定　　价：338.00元

---

# 序言 PREFACE

众所周知的是现在全球超过一半的人口居住在城市里，而且这一比例将继续增加。人类社会的发展带来了城市的兴起和人口的爆炸，现在我们的成功已经为我们赖以生存的自然系统带来了难以承受的压力。

城市是应对众多人口的最佳方案，城市里的资源和基础设施都可以共享，因而可以节约开放空间和自然系统。但城市并不是天生便可持续发展的。过去的一个世纪里，城市的兴起和扩张更多是事出偶然，没有经过规划。它们的框架，正如今天大多数城市所拥有的，并不一定支持我们理所当然认为的、城市该具备的功能。而且最近数十年，特别是在亚洲，都市越来越多的钢筋水泥、金属堆等已经威胁到个人的生存体验和身份标识。

因而我们在现代都市中遇到的挑战是如何重新规划城市蓝图，以解决这个全球性的问题而不是放任自流。我们知道这需要随着时间的推移而逐步完成，并不能一蹴而就。我们面临的挑战，如生态威胁、水污染、空气污染、水荒、气候改变、文化支离破碎、社会不公和公共基础设施需求。有些地方已经有了部分公共基础设施，有些地方的公共基础设施则更新、更好。因而我们所面临的挑战同时是环境性的、社会性的、技术性的和财政上的。

每个项目都有着特定的客户、选址和交流问题。我们认为，上述的是全球问题。全球各地都有我们的项目，它们都能在应对全球性挑战中起到相应的作用。这些不同地点的项目不再是孤军奋战的一员，而是成为了全球努力中的独特一员。全世界都在分享知识和经验，以打造更多彰显人类的聪明才智和责任的范例。

我们作为景观建筑事务所，在越大型的项目中扮演着越重要的角色。当景观建筑师被认为应考虑到土地使用的方方面面，我们便成为了应主导和推动项目的主要一方，而不只是在项目的最后修修补补。我们通过和规划者、工程师和建筑师积极合作和密切协作取得了成功。我们通过在区域性的项目中发挥正面作用，应对全球规模的挑战。

当交通规划者决定设置一条新的运输路线时，景观设计师会考虑到这样的基础设施硬件将如何与人类的地盘和生活融为一体。当环境科学家认定一条河流已经达到无法接受的污染水平时，景观设计师思考的则是修复后的自然环境如何能推动社会经济复兴。当城市规划者决定在一个地方兴建一个新的公园时，景观设计师便问人们想在这个空间内做什么和这个空间可以为它的都市和自然环境做何贡献。我们工作系统涉及的范围非常广，从水、土壤、植物，到混凝土和钢筋，到人，以及他们的信念和行为。我们的作用不是简单地执行设计任务，而是寻求可获得的、最广泛的效益。

杰西塔 · 麦肯
全球设计 + 规划 + 经济学领袖
AECOM 景观建筑事务所

It is now well known that more than half the world's people live in cities, and that this proportion will continue to increase. The rise of cities is an outgrowth of humanity's societal development, which has also produced a population explosion. Now our success has put unbearable pressure on the natural systems that sustain us.

Cities are the best solutions for large populations. Because in cities, resources and infrastructure can be shared while open space and natural systems are spared. But cities are not inherently sustainable. Cities have arisen and expanded more by happenstance than design in the last millennium. Their skeletons, such as they exist today, do not necessarily support the agendas we confidently see cities as representing. Furthermore, with the spike in urban growth of recent decades, particularly in Asia, the sheer weight of modern urban form threatens to overwhelm the experience and identity of the individual person.

So our challenge in dealing with the modern urban world is to reconfigure city blueprints so that they solve global challenges rather than perpetuating them. We know that this is not something done in a single sweep, but incrementally over time. The challenges we seek to address include: ecological threat, air and water pollution, water scarcity, climate change, cultural disintegration, social inequity, and the need for civil infrastructure – any in some places, newer and better in others. Our challenges are thus environmental, social, technical, and financial.

Each project addresses its own specific client, site, and community challenges, but we recognize that the aforementioned are global challenges. We practice globally so that our projects can each do their part in addressing global challenges. Rather than isolated expressions of locality, projects become locally unique patches of a global effort to share knowledge and experience and advance the paradigm of human ingenuity and responsibility.

We as landscape architects have been playing a bigger role in bigger projects. When the landscape architect is seen as someone responsible for considering all the interests that intersects in land usage, we become the perspective that should lead and drive major projects, rather than putting the finishing touches on them. We succeed by working proactively and collaboratively with planners, engineers, and architects. We are able to address global-scale challenges by bringing our perspective to the major projects affecting regional-scale systems.

When the transportation planner determines that a new transit line is needed, the landscape architect asks how this hard infrastructure will meld with human places and lives. When the environmental scientist determines that a river has reached unacceptable pollution levels, the landscape architect asks how restored natural environments can drive social and economic regeneration. When the city planner determines that space is available for a new park, the landscape architect asks what people want to do in this space and what this space can do for its urban and natural context. The large systems we work with thus range from water, soil, and plants, to concrete and steel, to people, their beliefs, and behaviors. Our role is not simply to execute a design task, but to imagine the widest benefit that can be achieved.

Jacinta McCann
Global Practice Leader, Design, Planning + Economics
AECOM.

# CONTENTS / 目录

## INTERVIEW/ 专访



## PROJECT / 项目

# OLIN

Landscape Architecture and Urban Design

# INTERVIEW / 专访

## Interview with OLIN

## 对话 OLIN

OLIN creates distinguished landscapes and urban designs worldwide. Our work is predicated upon social engagement, craft, detail, materiality and timelessness. Our appreciation of the urban environment is paramount. We love cities. We celebrate their complexities, contradictions and constraints. We believe in their continued evolution and enjoy collaboratively shaping the way they look and function.

Craft is seminal to our practice. Landscapes are intentioned to be aesthetically pleasing and durable. We value multi-functional, flexible and environmentally sensitive design solutions. Through craft, our landscapes have potential to be catalysts for human interaction, recreation and community. We approach each project individually, basing design decisions upon the underlying expressive power of a particular site in conjunction with specific programmatic requirements. In all of our work, we carefully assess the distinguishing spatial characteristics of a site, its connections to surrounding areas, local traditions and history, the natural environment and social and cultural conditions. We synthesize and integrate our findings in new site elements and systems that we manipulate, amplify and layer to create expressive designs.

OLIN's celebrated projects include Bryant Park, Canary Wharf, Battery Park City, the J. Paul Getty Center and the Barnes Foundation. OLIN's current work includes prominent work such as Philadelphia's Dilworth Plaza, the waterfront of Mill River in Stamford, Connecticut, and the Washington Monument Grounds at Sylvan Theater on the National Mall. We work on large-scale master planning commissions to mid-sized institutional work to small urban interventions. Regardless of scale or typology, our multi-disciplinary design staff create environmentally advanced, technical projects, promoting greater social engagement and ecology for every project.

OLIN卓越的景观和城市设计作品遍布世界各地，从社会参与、工艺、细节、物质性和非耗时性等都体现了我们的工作内容。我们认为对都市环境的审视欣赏必须被视为是最重要的。我们热爱城市，我们为这里的楼市建筑、矛盾冲撞和社会限制等而喝彩。我们相信城市会不断进化，并享受共同打造城市景观和城市功能的过程。

工艺对我们的实践至关重要。我们的目标是实现城市景观美学上的愉悦性和可持续性。我们重视具有多功能、灵活性的和环境敏感型的设计方案。通过人们的手工努力，我们的景观很可能会成为人类交流、娱乐和人类社区相融合的结晶。我们以基于具有内在表现力的城市景观和具体的项目要求来单独实施各个项目。在所有的工作中，我们都会对一处景观所具有的特色空间特征、它与周围景观的联系、当地的风俗习惯和历史人情、当地的自然环境和社会文化情况等进行认真仔细的评估。我们会把在一处新景观的构造和系统里的研究和发现融合到一起，然后我们会对这些新发现进行管理，放大和层层细分，以此来创造出令人印象深刻的景观设计。

OLIN已完成的著名项目有Bryant公园、Canary Wharf、Battery Park City、the J. Paul Getty Center 和 the Barnes Foundation等。OLIN目前正在实施的特色项目有费城Dilworth Plaza、Connecticut州斯坦福密尔河（Mill）水幕、国家购物中心Sylvan剧院华盛顿纪念碑广场等。我们的工作范围涉及面非常广，包含了大规模、大师级别的设计规划，到中型的机构性工作，再到小规模的都市改造。 除去规模，我们多才多艺的设计人员同时致力于完成有品质的、具有高技术含量的环境友好型项目，以此来提高每一个工程的社会融合度，实现生态平衡。

**A: Each company has its own culture, so what's the enterprise culture and mission statement of OLIN?**

A: 每个公司都有自己的文化，那么OLIN的企业文化和宗旨是什么呢?

**O:** Landscape requires stewardship. The field of landscape architecture must address some of the globe's greatest challenges: a compromised ecology, an aging infrastructure and the pressure to house growing populations. By orchestrating the complexities of the modern landscape, OLIN's designs promote community building and inspire engagement with the natural world.

OLIN is dedicated to affecting positive change through landscape architecture, urban design and planning. We are advocates for the artful creation and transformation of the public realm, and practice in a range of scales, including ecological and regional systems, urban districts, campuses, civic parks, plazas, and intimate gardens.

From our studios in Philadelphia and Los Angeles, OLIN crafts timeless spaces that promote social interaction and enhance life. We successfully realize projects locally and internationally, each one reflecting its unique context. Through rigorous research, analysis and a dynamic design process, OLIN incorporates the intrinsic qualities of a site to generate a landscape that is embraced by its community. Sustainability is a central tenet of our holistic approach, uniting natural processes with technical innovation to produce contemporary and beautiful places.

We collaborate with clients, architects, engineers, artists and communities to build consensus among diverse interests while attaining the highest level of design integrity.

O: 景观需要管理。景观建设领域必须应对全球一些最大的挑战，如受破坏的生态环境、老化的基础设施和人口增长形势下的住房压力。OLIN 的设计通过统筹现代景观建筑群，优化了社区建设并鼓舞了人们与自然的对话。

OLIN 公司致力于在景观建筑、城市设计和规划这几方面带动积极的改变。我们主张公共空间进行艺术性创新和改造以及各种规模的实践，包括：生态区域系统、城区、校园、市民公园、广场和私人园林。不管 OLIN 公司在费城的工作室还是洛杉矶的工作室都致力于设计永不过时的空间，以促进社会互动，提高生活水平。我们成功地意识到本土项目和国际项目都反映了该地区的独特环境。OLIN 通过严密的调查、分析和创新设计方案，将一个地区的固有性质组合起来以创造一个受当地人欢迎的景观。持续性是我们整体方案的一个中心原则，它将自然过程与技术革新综合起来以创造出现代而优美的景观。

我们与客户、建筑师、工程师、艺术家和社区合作，在不同利益之间达成共识，并且获取最高水平的整体设计。

**A: It has been nearly 40 years since OLIN was founded, and OLIN had carried out and finished many projects in this period. We know team cooperation is one of the necessary elements to achieve the success. So would you tell us that beside work time, how do OLIN enhance the group cohesiveness in spare time?**

A: OLIN 成立了将近 40 年，在此期间已经开展并完成了很多项目。我们知道团队合作是取得成功的要素之一。那么您愿意告诉我们在工作之余 OLIN 是怎样加强团体凝聚力的吗?

**O:** The life of the OLIN community extends well beyond regular project work and the office walls. OLIN has fielded a softball team in the local architect's league for Over 10 years, but in fact much of our social activity still revolves around our passion for design and cities. We regularly take part in pro bono design work through the Community Design Collaborative, a local organization that provides preliminary design services to non-profits and communities. Our closeness to the local community includes our participation in National (Park)ing Day, where we have participated the last 3 years by reclaiming a city parking space with a design produced by the office's staff in Philadelphia and Los Angeles.

OLIN was founded by two academics, so discussion has always been part of the culture. We go to lectures and exhibitions, and we talk about them over dinner or drinks. But as we've grown, we've needed to find more formal ways of incorporating it into the studio. For the last couple years, we've run a monthly, in-house theory symposium that gives us a venue to step outside the minutiae of practice and talk together about the ideas at the center of our work. Sometimes it's is as simple as a happy hour or brown bag lunch to present recent work. This means that you know what people are working on, and the next time you see somebody in the hall or the kitchen, you can talk with them about their work.

O: OLIN 团体一直发展得很好，不受项目工作和办公室的限制。我们在当地建筑师联盟里成立了一个超过 10 年的垒球队，但实际上我们很多社交活动仍然是基于对设计和城市的热情。我们定期通过 Community Design Collaborative 参加公益设计工作，它是一个为非营利机构和社区提供初步设计服务的当地组织。我们与当地社区的密切关系表现在我们在过去的三年里都参加了 National (Park) ing Day 活动，开发了一个城市停车场，由费城和洛杉矶工作室的员工设计。

OLIN 是由两位专业学者创办的，因此讨论一直是我们公司文化的一部分。我们去参加讲座和展览会，不管是用餐或喝酒都会进行讨论。但由于公司已经发展成熟起来，我们需要寻找更多正式途径将讨论合并到工作中。在过去的几年中我们通过每月举行一次内部理论研讨会，提供了一个集合点，这使我们走出繁琐的工作，能在工作地的中心交换意见，有时候陈述最近的工作成为了一件简单而快乐的事情，或是像展示简易午餐一样。这就意味着你可以了解大家在忙什么工作，下次在大厅或者厨房里见到某个人，你就可以跟他们聊最近的工作。

**A: Environmental issue is always the world's main concern. Especially issues such as global warming and air pollution, they have become more and more severe in these years. From the end of 2013 to the beginning of 2014, severe haze event occurs in east-central China, which has seriously affected people's daily life. Meanwhile, it has caused a certain amount of international influence and has become the focus of widespread attention. And what's your opinion about this issue?**

A: 环境问题一直是世界关注的焦点，尤其是全球变暖和空气污染，近几年已经成为越来越严峻的问题。从 2013 年末到 2014 年初，中国中东部发生了重度雾霾事件，严重影响了人们的日常生活。同时，它还造成了一定的国际影响，成为广泛关注的焦点。您对此有什么看法吗?

**O:** Greenhouse gasses and climate change are issues that the people of Earth face together. Political bickering and inaction does not improve our situation. What matters is that we do something meaningful and actionable. While the East considers basic environmental protection measures such as regulating pollutants from industry, in the West we are building on a legacy of environmental protection. Inherently, landscape design will improve air and environmental quality through increased use of vegetation which can sequester carbon and improve the air. Water can be purified by filtration through landscapes. In a very real way, the places we create provide much needed ecosystem services to help repair our planet, one project at a time.

O: 温室气体和气候变化是地球上所有人都面临的问题。政治争执和不作为并没有改善我们的情况。现在重要的是做些有意义的可行的事情。当东方人考虑采用基础性的环境保护措施，如控制工业污染物的排放等时，我们西方人则是依赖环保的传统。景观设计本身就是提高环境和空气质量的，它是通过增加植被来吸收碳化物，净化空气。过滤性景观可以净化水质。事实上我们创造的场所可以提供急需的生态系统服务，帮助修复地球。

**A: Sustainable developmental strategy is the long-term developmental guidelines of the whole world. As a famous landscape design company, OLIN has made a great contribution to many cities' development and has brought better living environment to more and more people. So the question is: how do OLIN implement the sustainable developmental strategy into each project?**

A: 可持续性发展战略已经成为全球性的长期发展指导方针，而 OLIN 作为一家世界著名的景观设计公司，已经为许多城市的发展做出了重大贡献，为越来越多的人创造了更好的生活环境。因此这里有一个疑问：OLIN 公司是怎样将可持续性发展战略实施到每个项目中的?

**O:** OLIN has a long legacy of designing projects that respond to and improve site living conditions and ecosystems performance. Our projects help to regenerate urban spaces for people and ecology. By designing with a deep appreciation for people along with nature, OLIN creates vibrant places that—of all types—from soil microbes, flora and fauna, and people alike. Sustainable design is an underpinning to our work, and each project in the urban setting is a unique opportunity to improve the performance of the landscape compared to typical anthropocentric values alone. We seek measurable and definitive metrics that define sustainability, including impacts to water, soil, air, vegetation as well as social metric—that indicate the creation of successful social spaces. Our most sustainable projects provide valuable ecosystems services while providing societal benefits.

O: OLIN 有着悠久的历史，设计了许多符合并能提高当地生活状况和生态能力的项目。我们的项目旨在为人们和生态革新城市空间。OLIN 的设计很大程度上得益于人们和自然，我们设计的场所充满生机，从土壤微生物、动植物群到人类各种类型的生活都可以得到改善。可持续性设计是我们工作的基础，在城市背景中每个项目与典型的以人类为中心的价值观相比，都是一个独特的优化景观的机会。我们寻找能够实现可持续发展（包括对水资源、土壤及社会指标的影响）的可行的决定性方案。我们公司最具有可持续性的项目提供了宝贵的生态系统服务，并带来社会效益。

**A: On February 22, Ben Monette (Senior Landscape Architect, OLIN) took part in a meeting which the theme was Integrating Natural Processes into the City Fabric of the Global American South at the Cities, Rivers, and Cultures of Change: Rethinking and Restoring the Environments of the Global American South Conference, which was presented by UNC Center for Global Initiatives. There were two environmental issues referred in this meeting: (1) efforts to restore natural and built environments, and (2) the implications and connections between changes to the American South and the inter-connected global environment in which we live. We'd like to know what new solutions or new ideas had been put forward during this two-day's long meeting?**

A: OLIN 的高级景观建筑师 Ben Monette 在 2 月 22 日参加了一个会议，主题为将自然过程结合到整个美国南部的城市肌理中，如城市、河流等，并且全球计划 UNC 中心提出了应反思和恢复 Global American South Conference 涉及的环境。该会议提到两点：（1）恢复自然、建设环境的工作；（2）美国南部变迁的隐含意义和联系，以及我们生活在的相互关联的全球环境。我们想知道这次为期两天的会议提出了哪些新方案或新观念呢？

**O:** The American South has become a key barometer for the nation and even the world in terms of environmental, economic, and socio-cultural evolution of cities and societies. With many northern U.S. cities suffering from population decline and aging infrastructure, the South in many ways offers opportunities for future-building. You can see this in cities like Austin and Raleigh, with active, young populations. You can see it in the unprecedented influx of start-up companies and high tech jobs within the last ten years.

With all of this momentum, the South is poised, like no other place in the U.S., to tackle the challenges that are facing many cities in the world, such as how to address outmoded infrastructure, and how to become more environmentally resilient in the phase of rapid population growth, drastic climate change, and limited government budgets. And one issue that touches all of these large questions is water management—how to conserve, direct, store, treat, and distribute this resource in a way that is at once economically viable, environmentally responsible, and culturally valuable.

The 2014 Global American South Conference brought together experts of diverse backgrounds to address these tough questions. OLIN Senior Landscape Architect Ben Monette presented his experience working in Cleveland, Ohio on a project called Green Over Gray. As a Rust Belt city, Cleveland has been going through many of the struggles of population loss, low investment, and aging infrastructure faced by so many post-industrial metropolitan areas. When Cleveland was mandated by the U.S. Environmental Protection agency to upgrade its obsolete combined sewer overflow system, the city decided to use their predicament as an opportunity to identify projects that would not only meet the mandate but serve as visible public investments in the city's urban realm These projects would then engage citizens, businesses, cultural and academic institutions and more to work toward a unified vision of Cleveland's future. The leaders of this initiative engaged OLIN to help visualize and prioritize a roadmap of catalytic projects that will guide the city in this expansive re-investment integration of environmental and socio-economic goals.

Specific research into engineered ecology combined Emily McCoy's presentation of the latest in urban site performance monitoring at the U Penn campus in Philadelphia with Dr. Margaret Palmer's research into the ecosystem function and desirable services support of constructed stream restoration.

In the end, all of the conference's discussions seemed to coalesce around a single imperative: the need for local action and advocacy. When governments become bogged down in red tape or sidelined by budget constraints, citizens, businesses, and academics can work together to keep the momentum going, united in the common benefit that is an environmentally resilient and future-oriented city. Ultimately the idea of an ecological stewardship model for city growth presents us with a foundation for a measurable and logical approach to the often diametrically opposed concepts of urban development and ecological restoration.

O: 美国南部在环境、经济、以及城市社会文化发展方面，已经成为美国甚至是全世界的一个重要的晴雨表。许多北部城市出现了人口减少和公共设施老化这些问题，南部在很多方面为未来建设提供了很多机会。你可以从奥斯汀和罗利这些人口活跃而年轻的城市中了解到这一点。你也可以从过去十年中创业公司和高科技就业史无前例的热潮中看出这一点。

在这样的趋势中，南部与国内其他城市不一样，依旧坚定不移地处理世界各地正面临的挑战，如：怎样解决过时的基础设施；怎样在人口迅速增长、气候急剧变化和有限的政府预算这些条件下增强环境的弹性。与这些大问题都有联系的一点是水资源的管理，即如何以一种经济上可行、对环境负责且有文化价值的方式来保护、引导、储存、处理和分配这些资源。

2014 年美国南部全球会议召开期间，来自不同背景的专家汇聚一堂来解决这些棘手的问题。OLIN 高级景观建筑师 Ben Monette 分享了在克利夫兰市一个名为“Green Over Gray”的项目的经验。克利夫兰被称为锈带城市，一直存在着人口减少、低投资和公共设施老化等很多后现代工业化都市圈面临的问题。当克利夫兰应美国环境保护署要求改造废旧的合流下水道的溢流系统时，他们决定把窘境变为机会，来促成一系列既符合美国环境保护署的要求、又能成为城市公共区域投资的项目。这些项目会吸引市民、商业、文化以及学术机构的参与，更会促进克利夫兰未来的一致的愿景的实现。这次提议的领导者雇用 OLIN 公司协助设想及制定这些起催化作用的项目的方案，将在这次综合考虑环境和社会经济的大规模投资引导城市走向。

具体的工程生态学研究将 Emily McCoy 女士在费城宾夕法尼亚州大学校园所做的关于城市基地性能监控的最新陈述与 Margaret Palmer 博士关于生态系统的功能和如何较好恢复人工流的研究结合起来。

最后，所有的会议讨论似乎围绕着一个中心问题：本土作为和支持的必要性。当政府陷入官僚主义繁文缛节的泥沼之中或预算受到限制的时候，市民、商人和学者能够一起合作继续支持这些项目，为了城市和环境更有弹性更好的未来一起努力。最后，关于城市成长的生态管理模式的想法向我们展示了一个可行的逻辑性的方案根据，以处理那些与城市发展和生态恢复截然相反的观念。

**A: In OLIN's blog, there is an article written by Dalia Zein and was originally published on Landscape Architecture Network. called "10 Things You Must Know If You Want to Study Landscape Architecture", which provides a wonderful study way to the landscape design students. Would you like to tell us the necessary qualities to be one of the members of OLIN?**

A: OLIN 公司博客上有篇文章名为“研究景观建筑必须知道的 10 件事”，这为景观设计的学生提供了一个极好的学习机会。您愿意跟我们说说成为 OLIN 一员的必备条件吗?

**O:** Passion; Attention to details — Noticing the intricacies and sweating the small stuff is a necessary evil of being a landscape architect within OLIN. No detail is too small to go unnoticed, from the tallest of trees to the smallest of grasses, members of OLIN's studio take pride in every detail and aspect of a design. Artistic Eyes; Thought leader — Setting the pace is an important tenant of OLIN's philosophy. Through innovation and originality, the spaces created are in a world of their own, something that could not be created by 'following the trend'. Take initiative; Team player — Creating great designs is a collaborative effort, with nearly 80 individuals in OLIN's studio, working together is vital. Motivation; Communicator; Diverse interests — Diversity leads to great ideas. Multiple viewpoints and contrasting interests foster a unique ethos that cultivates brilliant designs. Global vision — Our world is now connected more than ever, and the reach of OLIN's work is continually expanding.

O: 首先是激情。此外，注重细节，成为 OLIN 的景观建筑师的关键是留心项目的复杂之处和注重细节，没有什么细节是可以忽略的，从高大的树木到最不显眼的小草，OLIN 工作室的员工都以设计中的每一个细节和每一方面为傲。OLIN 公司理念中重要的一点还包括员工必须具有艺术家一样的眼光，能成为思想上的领袖，并能掌握进度。通过革新和创新，那些创造出来的空间将成为属于他们的独一无二的世界。此外，还需积极主动，有团队精神，因为 OLIN 伟大的设计是工作室将近 80 名员工努力合作的结果，合作对 OLIN 来说非常重要的。同时，动力、沟通、不同的兴趣和多样性将助于伟大的设计诞生，广泛的视角和多样的兴趣能形成独特的氛围，促成杰出的设计。最后一点，全球视野。因为随着全世界的联系越来越密切，我们的工作范围也在不断扩大。

**A: We have noticed that OLIN doesn't have many projects in China. We'd like to know whether you have any plans to expand Chinese market.**

A: 我们注意到 OLIN 公司并没有在中国开展很多项目，因此很想了解你们是否有计划开拓中国市场?

**O:** In past 20 years, China has shifted 300 million people into the cities. The size and the speed of the urbanization are unprecedented. Its impacts on all aspects of global affairs, economics, nature resources and environment, and culture cannot be ignored, no matter it is positive or negative. Having a continuing healthier urban development and transforming the current cities to better living environment for Chinese citizen will benefit the whole world. With our nearly 40 years place making experiences, the understanding of what takes to make a better urban landscape for people to live in, and the global vision, OLIN desires and is cable to contribute more in this next phase of Chinese urban development. We are expanding our presents in Chinese market currently in the form of increased numbers of Chinese projects in our office. We are still in the process searching an effective way to do more that not only serves our desire for large impactful projects, but also satisfying our larger goal and mission. We are continuing exploring other ways to make positive impact such as, advocating the importance of living system in urban landscape development, involve cultural and intellectual exchanges with Chinese practitioners and scholars in the landscape architecture field.

O: 在过去的 20 年里，中国有 3 亿人口涌入城市。这种城市化的规模和速度是史无前例的。它对全球的事务、经济、自然资源、环境和文化方面带来的影响，不论是积极的还是消极的，都不容忽视。持续而稳健的城市发展和市民生活环境的优化将造福整个世界。我们有三十年的打造空间的经验，了解如何创造一个宜人的城市景观，加上我们的全球视野和设计渴望，可以为中国下一个城市发展阶段作出更多贡献。我们正在扩大在中国市场中的参与工作，工作室已有越来越多的中国项目。我们仍在寻求一种有效的方法：不仅能为大型有影响力的项目服务，而且能满足我们更大的目标和使命。我们在不断探索其他有积极影响的方式，如倡议城市景观发展中生活体系的重要性，与中国开业者和学者们在景观设计领域的广泛文化和知识交流。

# Camana Bay

# 卡马纳海湾

Landscape Architect: OLIN – Landscape Architecture and Urban Design

Location: Grand Cayman, Cayman Islands

A unique tropical paradise is the setting for a 202-hectare new community consisting of four village areas: a hotel resort and beach club; a town center with retail and commercial office space; a marina village; and a collection of residential neighborhoods. The design sets a new standard of development in the Caribbean and puts sustainability at the forefront. As part of a multi-faceted design team, OLIN provided urban design and master planning from a landscape architectural perspective, helping to develop the goals and philosophy of the community in response to its distinctive natural environment. The plan takes advantage of Cayman's beautiful setting and gentle trade winds; the challenge came in designing for the island's harsher conditions, such as fierce hurricanes, strong sunlight, high humidity, frequent seasonal rain showers, and salt-laden soil, water, and air. By funneling cooling breezes, incorporating open spaces, and producing shade with architecture, fabric, and plants, a comfortable pedestrian environment was created.

一个独一无二的热带天堂，占地 202 公顷，是由四个乡村地区组成的新社区。这四个乡村分别是一个度假村兼沙滩俱乐部，一个包括零售店及商业办公室的市镇中心，一个滨海村以及一个集中居住区。这个设计制定了加勒比发展的新标准，并将可持续性放在了首要位置。作为多才多艺的设计团队，OLIN 提供了城市方面的设计和从景观建筑角度出发的总体规划，来协助目标的发展实施，加强社区人保护独特自然环境的观念。这项计划利用了开曼优美的环境和逐渐兴旺的贸易活动。海岛恶劣的条件给设计带来了挑战，例如猛烈的飓风、强烈的阳光、较高的湿度、频繁的雨季以及含盐的土壤、水和空气。漏斗式的冷风，一体化的开放空间，建筑物和植物带来的树荫，这些创造出舒适的步行环境。

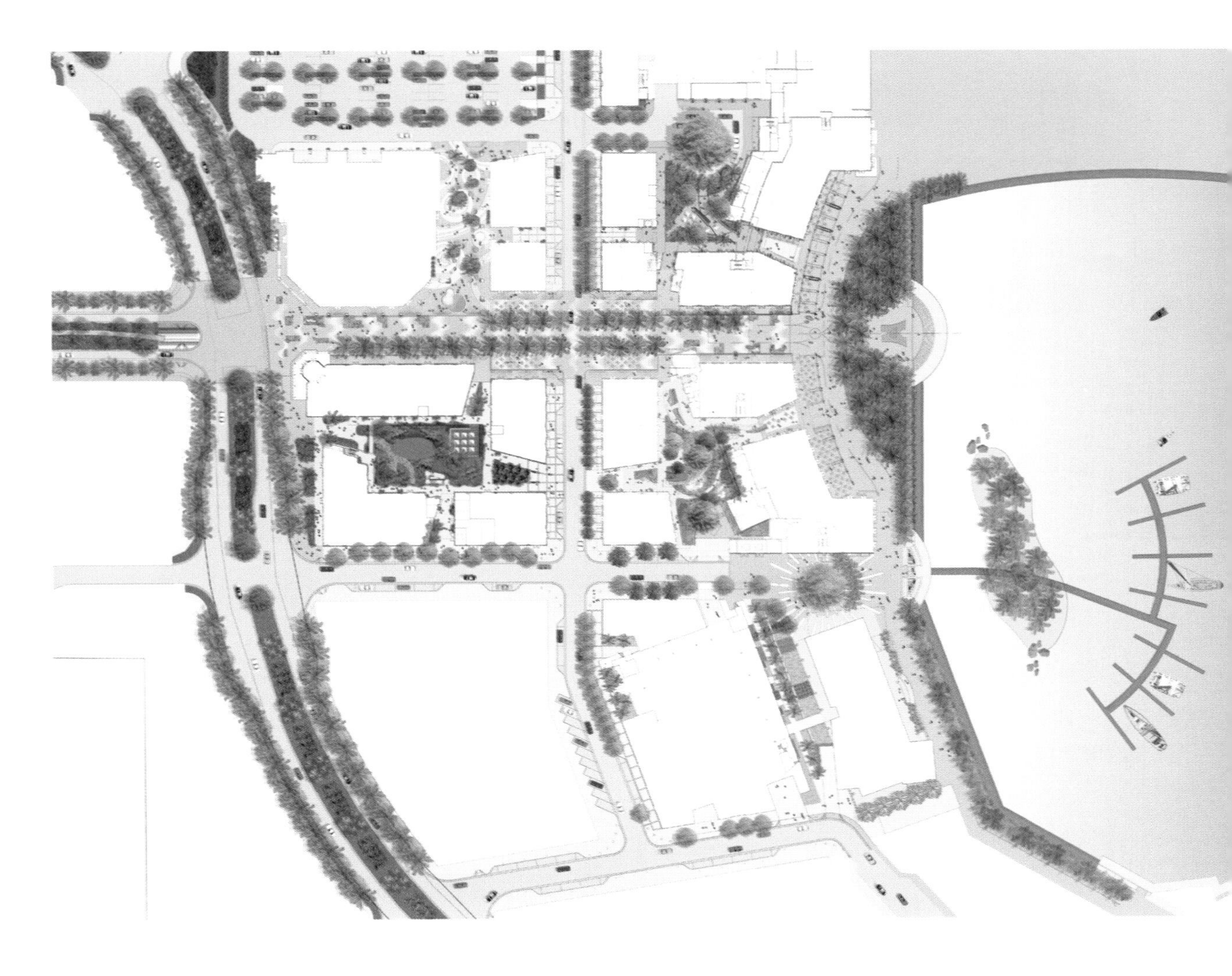

Camana Bay Plan

# Mill River Park and Greenway

## 磨河公园和林荫路景观

Landscape Architect: OLIN – Landscape Architecture and Urban Design

Location: Stamford, CT, USA

OLIN created a master plan with a layered, diverse and sustainable approach to programming the park. There are four distinct themes to the program that focus on the river: natural/ecological, cultural/educational, recreational, and experiential. The park will bring an active, vibrant and verdant environment to Stamford's downtown riverfront, while connecting adjacent neighborhoods to the river, making it a prominent, alluring and accessible place for the entire community. Paths along the river support walking and jogging, as well as provide multiple venues for resting and viewing. A carousel, café, fountain and ice skating rink, to be implemented at later phases will offer year-round leisure, entertainment, and recreational opportunities, promoting a festive, social atmosphere. The planting palette along the river's edges and slopes is durable and capable of withstanding storm events. The native plants, including grasses, wildflowers, shrubs, understory trees and canopy trees, are carefully coordinated to revitalize degraded aquatic and terrestrial habitats. Additionally, the resultant habitats will become an educational opportunity for both students and the amateur birdwatcher alike.

OLIN 制定了蓝图，对公园进行分层次、多样化、可持续性的规划。河流规划方面有 4 个显著的主题：自然 / 生态、文化 / 教育、娱乐、体验。公园会在斯坦福德市中心的河边地区营造出一个活跃、充满生机的绿化环境。同时它将邻居们与河流联系起来，使之成为一个显著的、充满吸引力又能让整个社区接近的场所。河岸边的道路可供散步和慢跑，同时为路人的休息和欣赏风景提供了多样化的场地。后期会增加旋转木马、咖啡馆、喷泉、溜冰场这些设施，提供全年的休闲、娱乐和消遣，增加节庆和社会活动的氛围。河岸及斜坡上有种植带，能持久抵御暴雨的侵袭。本土植物包括草地、野花、灌木、下层木、树冠木，它们被精心布局用来恢复那些已经退化了的水陆栖息地。另外，这些合成的栖息地为学生和观鸟爱好者提供了观赏与研究的机会。

# Syracuse Connective Corridor

## 锡拉库扎连廊

Landscape Architect: OLIN – Landscape Architecture and Urban Design

Location: Syracuse, NY, USA

For decades, the City of Syracuse has been divided both physically and culturally. The large overpass of Interstate 81 bisected neighborhoods and impeded a cohesive urban identity. The design of the Connective Corridor is inspired by a renewed vision for Syracuse that will connect and enliven a central artery of the city in need of revitalization. A major initiative of the project is to transform the Corridor into a linear urban stage, a promenade for businesses and institutions to engage pedestrians with an inspirational, fulfilling and diverse experience. To emphasize Syracuse's identity and involve local industry, the design utilizes materials and manufacturers from the city and region, a move which also vastly reduces costs by using relatively simple, everyday materials in new and thought-provoking ways. The use of native plantings and porous recycled materials serve as stormwater filtration devices and provide an educational opportunity along the Corridor. Through the innovative use of locally-sourced, common materials, the Connective Corridor demonstrates that low-cost, high-quality designs are possible, and that simple measures can have a profound effect upon a community's civic identity.

ONE WAY

PARKS

Syracuse Connective Corridor Plan

几十年来锡拉库扎市在地域分布方面和文化认识方面都有了分歧。81号州际公路的大天桥不但将社区一分为二，并且阻碍了城市凝聚力的发展。连廊的设计源于对锡拉库扎的一个新思考，即需要复兴式地连接并活跃城市的中央主干道。该项目的一个主要提议是将连廊转变为一个线性城市舞台，同时又是分布了商业机构和研究院的广场。带给路人富于启示性且伴随着成就感的多样化体验。 为了提高锡拉库扎的知名度，发展当地的工业，该项设计利用了城市各个区域的制造商提供的材料。用这种发人深思的新方法来使用简单的日常材料，也极大地减少了成本。本地植被和多孔回收材料被用作雨水过滤设备，同时也给连廊周边提供了学习研究的机会。通过对本土普通材料的使用，连廊证明了低成本，高质量的设计是能够实现的，同时也证明了简单的措施能对社区公民的认可产生深刻的影响。

# The Barnes Foundation

## 巴恩斯基金会

Landscape Architect: OLIN – Landscape Architecture and Urban Design

Location: Syracuse, NY, USA

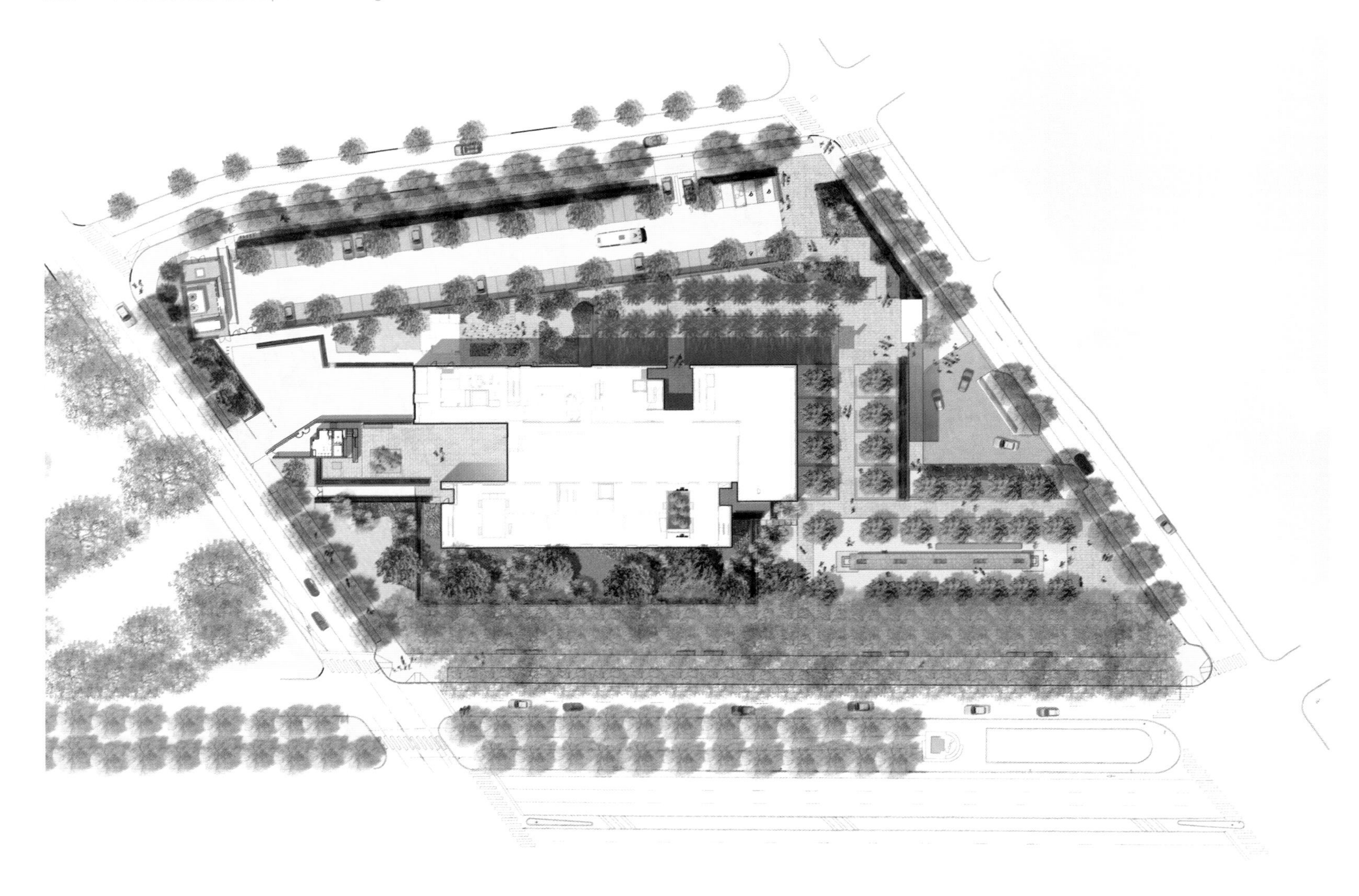

The Barnes Foundation Plan

For decades, the iconic art collection assembled by Dr. Albert Barnes was displayed at his summer estate turned museum and arboretum in Merion, PA. Now relocated along Philadelphia's Museum Mile, the priceless collection is accessible to the public as never before. Conceived as a gallery within a garden and a garden within a gallery, the design honors the original Barnes estate designed by Paul Philippe Cret, and provides visitors with a highly personal and contemplative experience. OLIN's design captures the spirit of the original site and fits into the urban context and implements contemporary sustainable practices. The gallery building opens to two separate outdoor spaces — one a lush garden and terrace adjacent to the museum's café, replete with plantings characteristic of the historic gardens in Merion. The other is a paved open terrace overlooking the parkway, suitable for more intimate events, decorated with antique benches, platforms for seating, loose furniture, a fireplace and a collection of flowering trees. Within the gallery a central courtyard has been left open to the sky. Graceful allee of trees surround the site and line the entranceways along with linear pools. Vine-covered walls and hedges discretely mask the visitor parking and bus loading points. An underground cistern captures stormwater for on-site irrigation, conserving resources and contributing to the project's LEED® Silver certification goals.

几十年来，由阿尔伯特·巴恩斯博士收藏的标志性艺术品受到越来越多的人的关注，因此他的避暑庄园也逐渐发展成了费城梅里恩的博物馆和植物园。搬迁到费城的博物馆史无前例地开放给人们。作为“花园中的画廊，画廊中的花园”，这个设计为巴恩斯基金会增添了荣誉，也给游客带来宝贵的难忘的观赏体验。巴恩斯基金会最开始是由保罗·菲利皮·科莱特设计。OLIN 的设计抓住了原址的精髓，并将它融入到现代城市环境中，也实践了当代可持续发展的决策。画廊正门朝向有两处分开的户外空间。一处是苍翠繁茂的花园跟平台，与博物馆的咖啡屋相邻，里面各种植物密布，承载着梅里恩历史性花园里植物的特性。另一处是铺砌的开放式平台，俯瞰着公园道路，适合更隐私的活动。这里有古老的长凳和平台可以就座，还有分散的家具、壁炉和开满鲜花的树木。画廊中间有一处露天庭院，曼妙的林荫小道包围着这个景点，沿着直线型水池排成排标示出画廊的入口。覆满葡萄藤的墙壁和树篱分散地装饰了停车场和公交站点。地下有一个收集雨水的蓄水池，用于灌溉和资源保护，以及促成该项目获得 LEED 银级认证。

# Dunkin Donuts Plaza — Horizon Garden

## ▶ 唐恩都乐广场 —— 水平线庭园

Landscape Architect: Mikyoung Kim Design

Client: Rhode Island State Council on the Arts

Location: Providence, Rhode Island, USA

Area: 557 $m^2$

Photography: Mark LaRosa

This sculptural garden creates an art environment that layers paving, landform, customized seating, and stainless steel sculptural elements to create a comprehensive experience of fluidity and transformation of light. The arts concept utilizes light inspired by light at the horizon line to fill this shaded space during the day with a golden hued light — integrating "sunlight" into the space. Throughout the day, the golden hues change slowly and transform into orange colors at the base of the sculpture, gradually transforming into blue hue in the evening. The light colors will ripple through various hues of the same color.

The sinuous forms of the sculpture flow through the garden, defining pedestrian movement and creating spaces for intimate seating as well as larger gatherings. The design of the curvilinear paving pattern highlights the central role of the art element as a source of warmth, organization, and visual interest within the space. The paving pattern emanates from the curvilinear forms of the sculpture that frame the various sub spaces in this garden. Luminous recycled blue glass will be handled on the surface of the concrete to create these arcs.

The gesture of the paving design and the sculpture visually connects the garden to the street. The form of the sculptural landforms and plant placement frames the garden spaces and reinforces the fluid reading of the entire design. The garden and sculptural materials work together with the light to create an oasis during both day and night for individuals to engage and larger groups to gather.

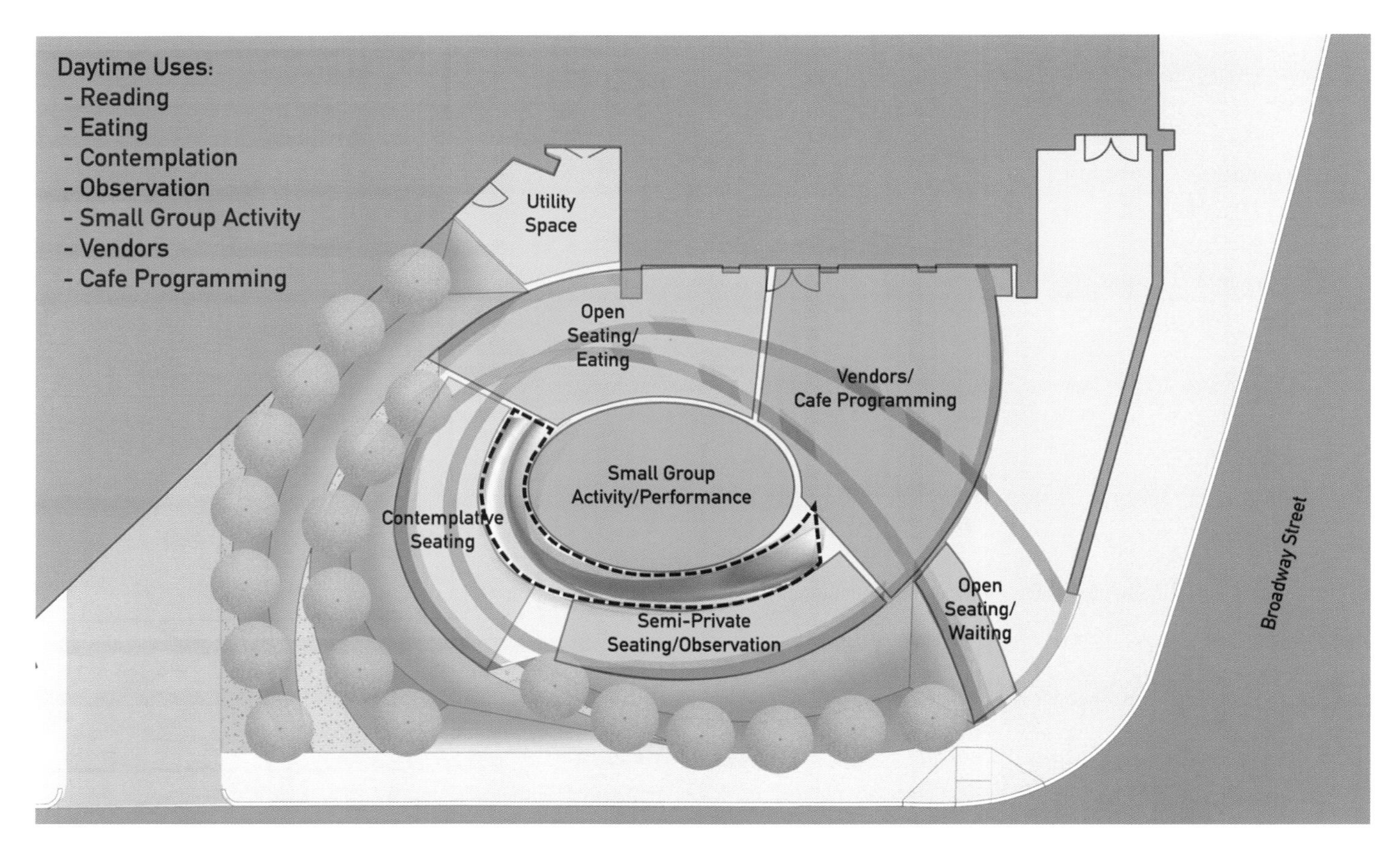


Programming Diagram - Day

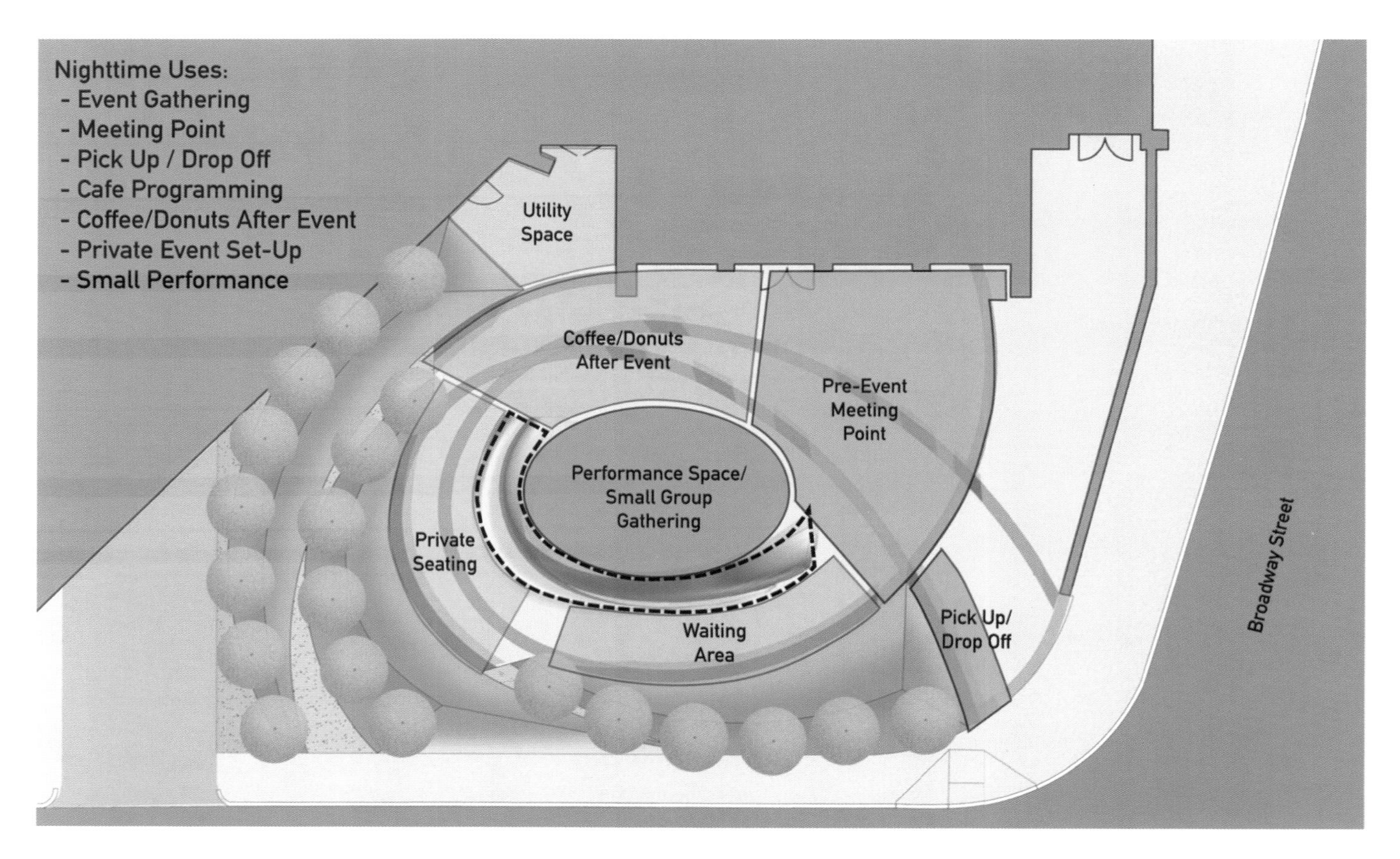


Programming Diagram - Night

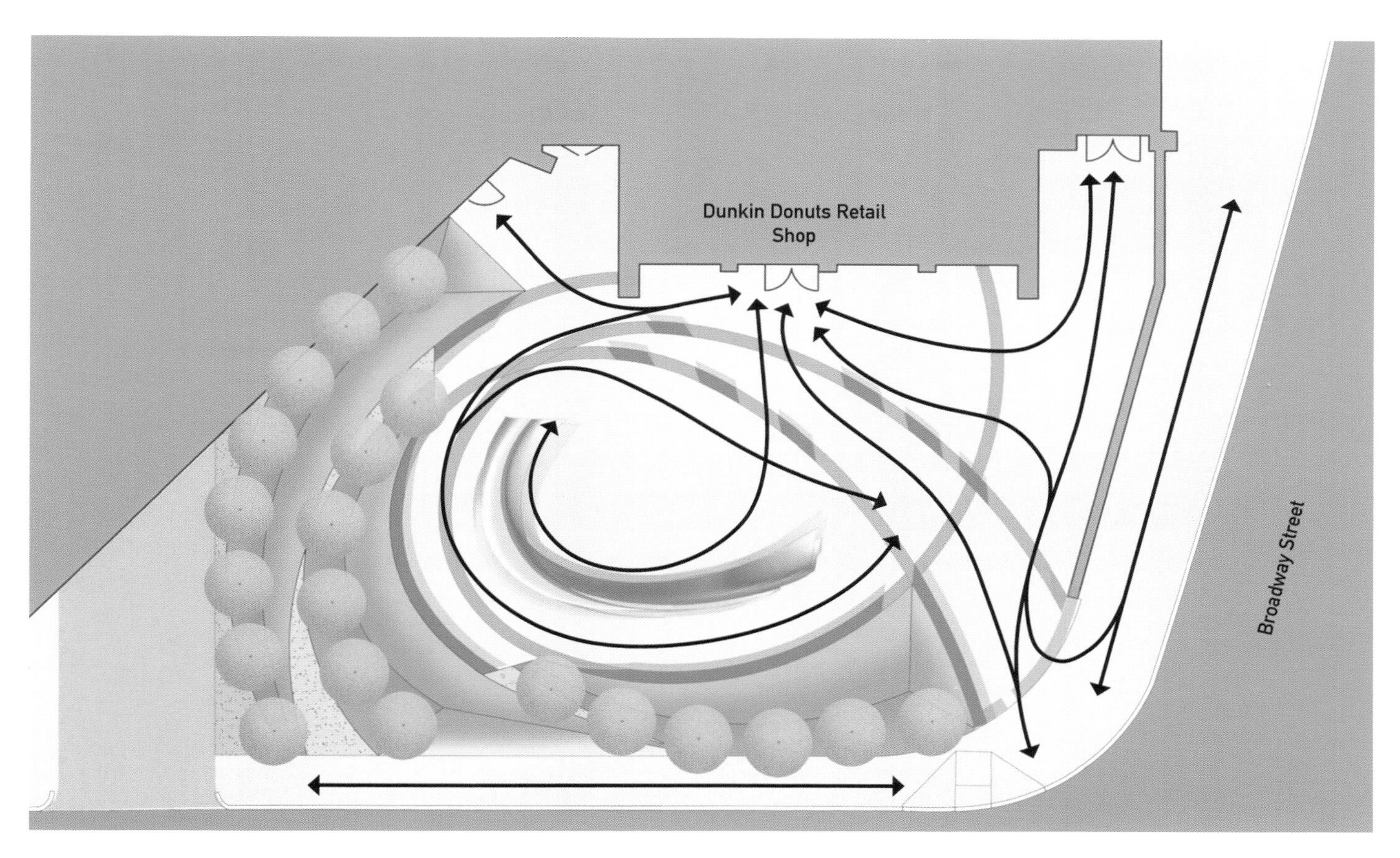


Programming Movement Diagram

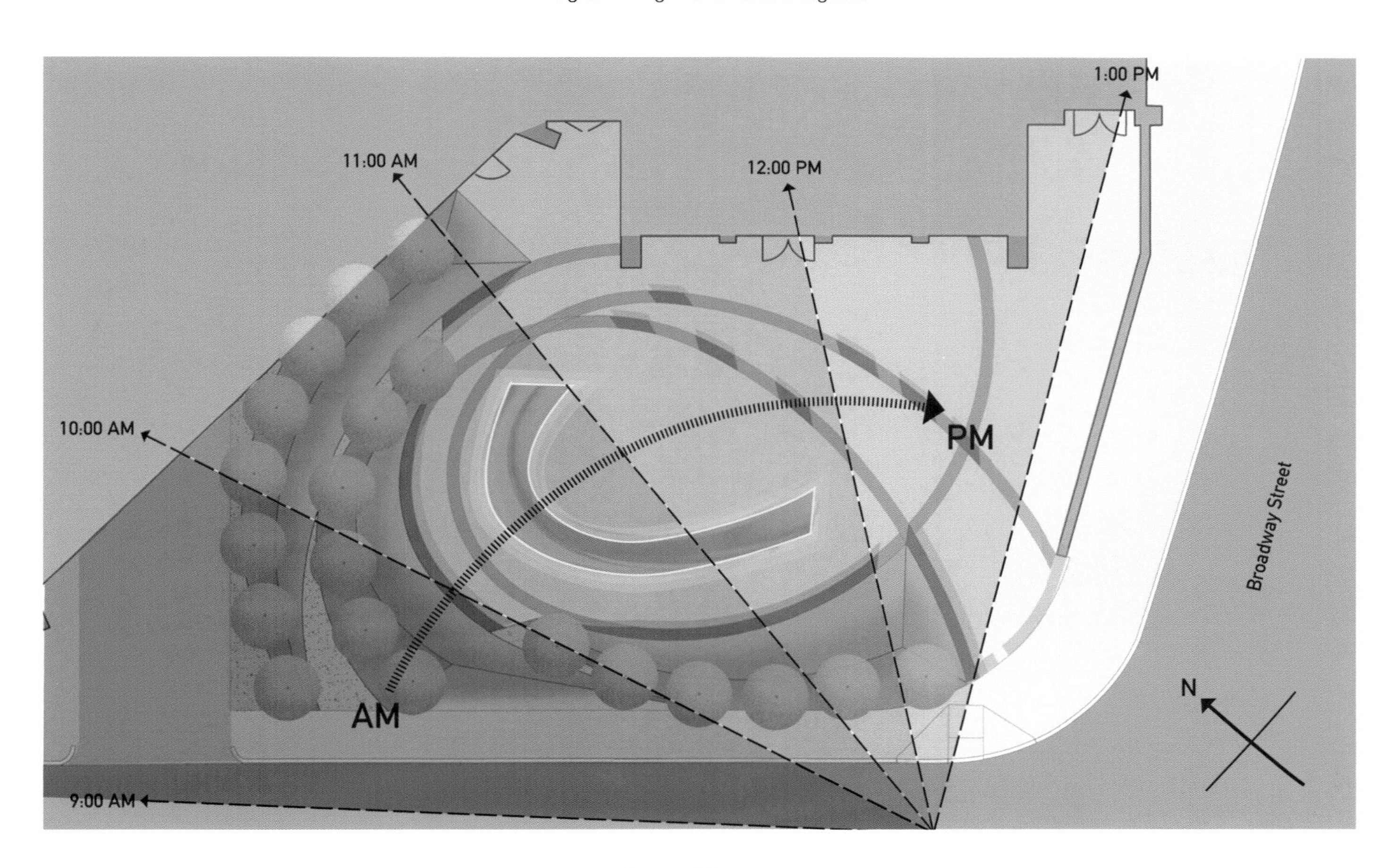


Shade Diagram

这个雕塑般的庭园创造出一个艺术环境，各种夹层的铺砌、地形、定制的座椅和不锈钢雕塑品让人全面体验到光的流动和变幻。白天利用金色的阳光来照亮这个阴影空间，这个艺术概念的灵感来源于地平线的光。一整天里，雕塑表面的光由金色逐渐变成橘色，到晚上渐变成蓝色。这些光的颜色因色调不同而形成涟漪。

一些雕塑以蜿蜒的形式穿过庭园，控制了行人流量，为私人休憩及大型聚会提供空间。铺砌的曲线图案突出了这里的艺术品作为热情、组织和视觉趣味来源的重要地位。图案起点是曲线形的雕塑品组合，规划出公园的子空间。混凝土表面铺的是循环利用的发光蓝色玻璃，用来制造弧线。

铺砌的样式和雕塑品从视觉上将庭园与街道连接起来。雕塑般的地形和植物的布局规划了庭园的空间，加强了整个设计的连贯性。庭园、雕塑材料和光的共同作用为个人游玩及大型团体聚会创造出一片绿洲。

# Giant Interactive Campus

Landscape Architect: SWA Group

Location: Shanghai, China

Area: 182,000 $m^2$

Photography:Tom Fox / SWA Group

The 182,000 $m^2$ campus and green roof for Giant Interactive Group in Shanghai, China, was conceived as an ecological park and living laboratory. Structured around a plan of natural systems and open space, both the landscape and architecture programs seamlessly integrate. Half of the campus contains corporate and office uses, while the other half is focused on lifestyle, and includes a hotel, clubhouse, along with dining and recreational spaces.

SWA Group sought to use the landscape as an organizing framework for the master plan with a variety of water experiences focused on ecological sustainability. Many of the existing trees were salvaged and relocated across the campus. A city road (Zhongkai Road) bisects the land, splitting it into two sections. Planning of the site employed water as a means to connect and organize various program elements. An intricate hydrological system consisting of existing irrigation canals, new water retention basins, islands, and seasonal wetlands, created a diverse habitat for wildlife as well as scenic points.

The green roof was designed to be a low maintenance "meadow" that requires little watering and naturalizes over time. Unlike a typical green roof, the surfaces fold, soar and dip. The undulating roof structure touches the ground plane, dipping into the adjacent wildlife pond as well as coming into contact with the pedestrian plaza.

The extreme slope conditions vary up to 53° and posed significant challenges for vegetation. An innovative system of reinforced concrete cleats, spanned by steel angles and gabions, are laid parallel to each sloping surface. The system functions as large self-contained cells holding the soil in place and thus minimizing slumping and erosion due to gravity. The roof's extensive size acts as a thermal mass that limits heat gain and reduces cooling expenditures.

Riparian plants, and a wetland sanctuary of networked marshes and islands make a comfortable buffer balancing work and lifestyle. The overall design outcome is the seamless connection of landscape, architecture and environment to the site.

位于这个 182 000 $m^2$ 的园区的中国上海巨人集团总部园区及绿化屋顶是一个生态公园和生活实验室。该设计将景观和建筑完美地结合在自然系统与开放空间中。园区一半包含公司及办公室设施，另一半则是生活休闲区，餐饮娱乐区旁设有一家酒店和俱乐部。

SWA 试图在景观中设计以生态可持续性为重点的水景，作为总体规划的一个组织性框架。许多现有的树木被修剪并迁移到园区里。中凯路将陆地一分为二。水域的设计是为了连接并组织各种项目元素，这个包括当下的灌溉渠、蓄水池、小岛和季节性湿地的复杂水系为野生生物创造出一个多样化栖息地，也开发出了很多景点。

绿化屋顶这片“草地”易于养护，只需要少量浇水和移植。它的表面是折叠的，起伏不平，与典型的绿化屋顶不同。这种波状的屋顶结构触及平地，融入邻近的野生生物池塘并与人行广场相连接。

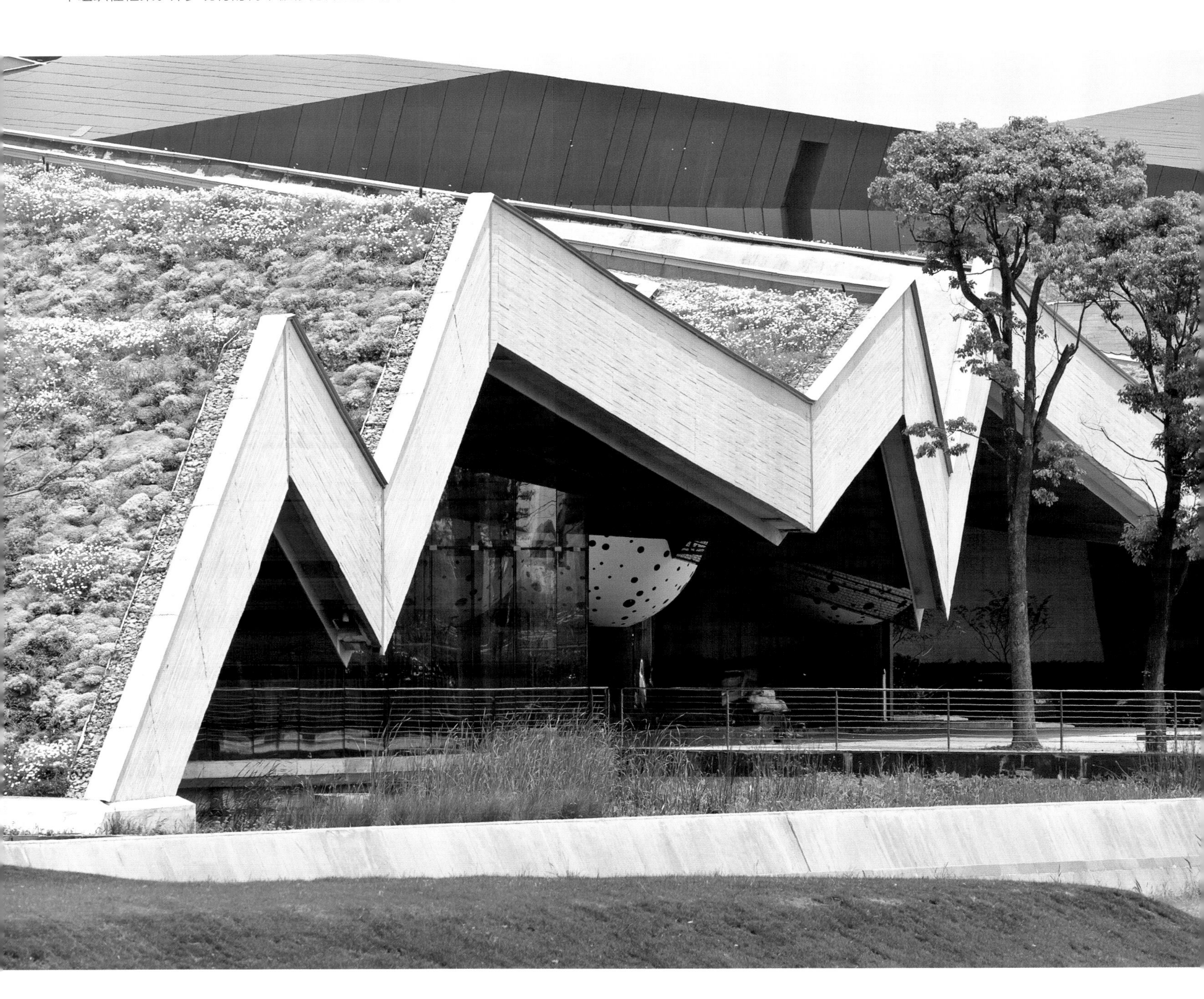

屋顶的极端坡度达到 53°，这为植物种植带来挑战。因此设计师设计了角钢和石笼固定的创新性钢筋混凝土夹板系统，并将其放置在每个倾斜表面的平行处。这个系统作为设备齐全的隔间将土壤固定起来，把重力造成的流失和侵蚀降低到最小。广阔的屋顶像是一个限制热量吸收并减少降温成本的温度调节体。

水边的植物、分布着网格状沼泽地和群岛的湿地保护区，这些使工作区与生活区达到舒缓的平衡状态。总体设计将景观、建筑和环境完美地与当地相结合。

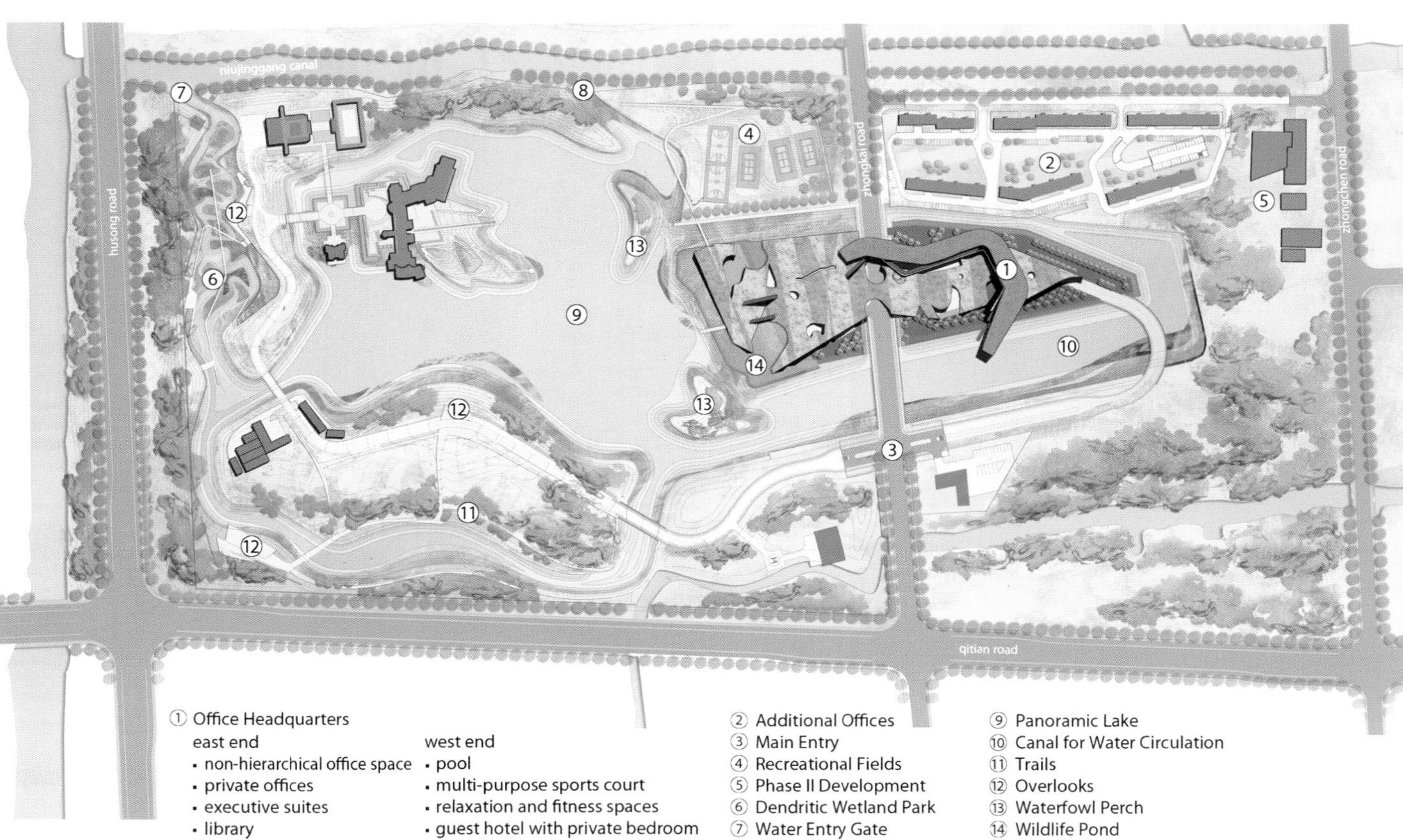


① Office Headquarters
east end
- non-hierarchical office space
- private offices
- executive suites
- library
- auditorium
- exhibition space
- cafe

west end
- pool
- multi-purpose sports court
- relaxation and fitness spaces
- guest hotel with private bedroom suites overlooking wildlife pond

② Additional Offices
③ Main Entry
④ Recreational Fields
⑤ Phase II Development
⑥ Dendritic Wetland Park
⑦ Water Entry Gate
⑧ Water Exit Gate
⑨ Panoramic Lake
⑩ Canal for Water Circulation
⑪ Trails
⑫ Overlooks
⑬ Waterfowl Perch
⑭ Wildlife Pond

Giant Interactive Campus Plan

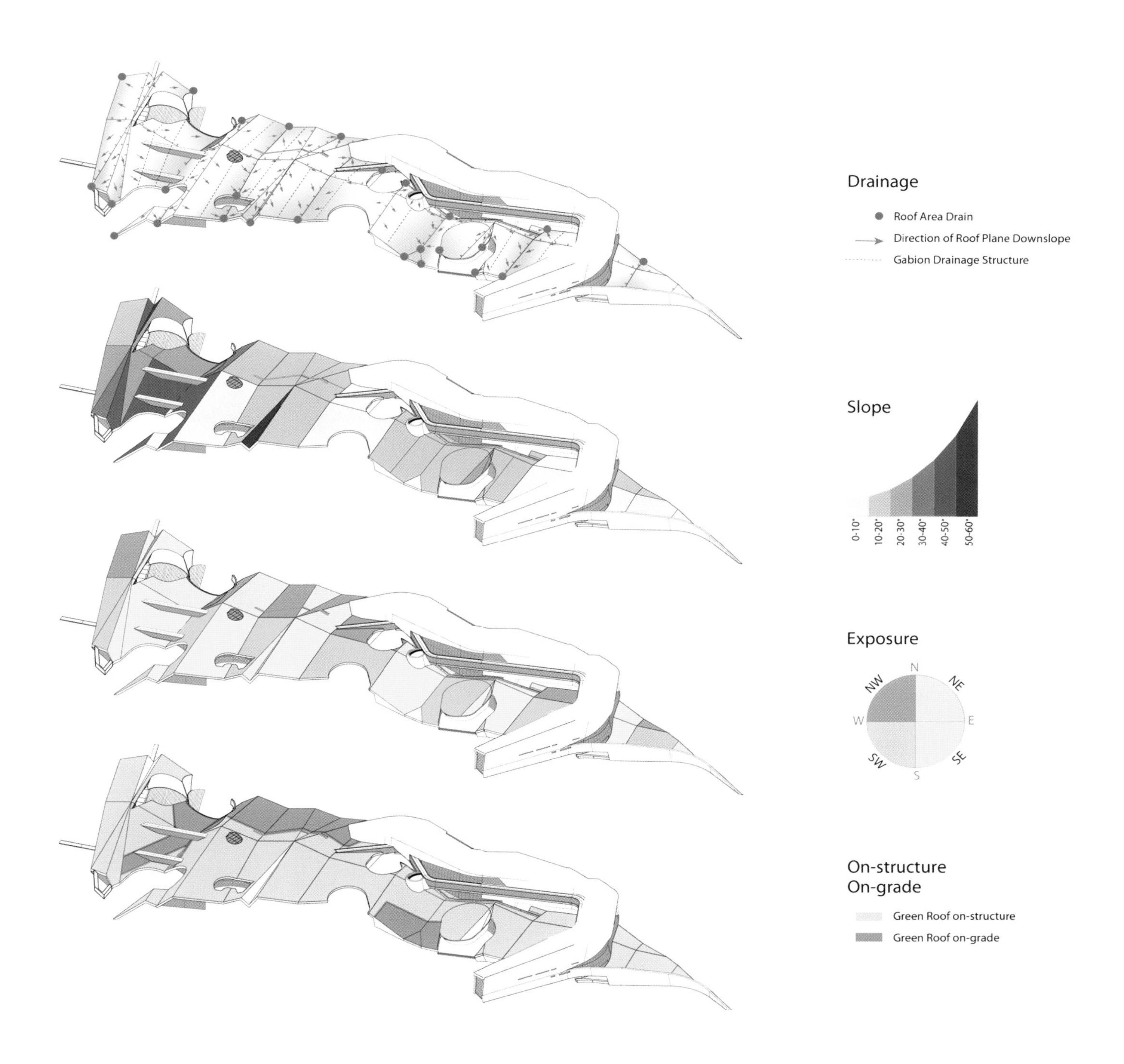


Landscape systems diagrams. Variation in drainage, slope, exposure and soil depth contribute to the overall performance of the Giant green roof.

Giant Interactive Campus Landscape Systems Diagrams

# Museo Del Acero Horno$^3$

## 阿塞罗奥尔诺博物馆

Landscaped Architect: Surfacedesign Inc. + Harari arquitectos

Design Team: James A. Lord, Claudia Harari, Geoff Di Girolamo, Roderick Wyllie

Client: Mr. Luis López, General Director of the Museo Del Acero Horno3

Location: Monterrey, Mexico

Photography: Paul Riveria / Archphoto, Abigail Guzman Tamex/ Grafix, James Lord

The overall landscape design emphasizes the physical profile of the 70 m furnace structure while complementing the modern design of the new structures. Large, free-formed steel objects and machinery unearthed during excavation are incorporated as stepping stones and other features. The design approach melds industrial site reclamation — and the adaptive re-use of on-site materials — with ecological restoration through the use of green technologies.

Two water features are integral to the narrative of the project, while helping to define and locate the public space adjacent to the museum. In the main esplanade, the steel plates that formerly clad the exterior of the main hall were repurposed into a stepped canal over which water cascades. The 200 m-long feature alludes to the tracks used daily to train in the thousands of tons of raw materials that are off-loaded in this location, and serves as a visual connection to the rain garden in the landscape beyond. At the museum's entrance, the stepped canal culminates in the misting fountain, a grid of rocks visibly embedded with ore. This trompe l'oeil evokes the caustic heating process once used to extract ore, but instead of steam it generates a cooling mist that blows over the plaza — a pleasant surprise for visitors in Monterrey's hot and arid climate.

The use of green roofs (extensive and intensive) over the museum which comprises the largest such roof system in Latin America — helps to reduce the visual impact of the new buildings. The existing furnace rises from this newly created ground plane. On the higher roof, a variety of drought-tolerant sedums have been arranged according to the structural roof patterns of the new architecture, and are contained by what appears to be a floating steel disk. A circular viewing deck allows visitors to take in the expanse of surrounding regional landscape, including the distant Sierra Madres, which are echoed in the roof's mounded shape.

Below, Alfombra verde (green blanket) a less constrained meadow of tall grasses — an abstraction of the native landscape — creates a connection to the landscape's pre-industrial context both functioning as a bioremediation for degraded soil and increasing thermal benefits for the new structure.

Principals of sustainability are at the core of the landscape design of the Museo Del Acero Horno[3].

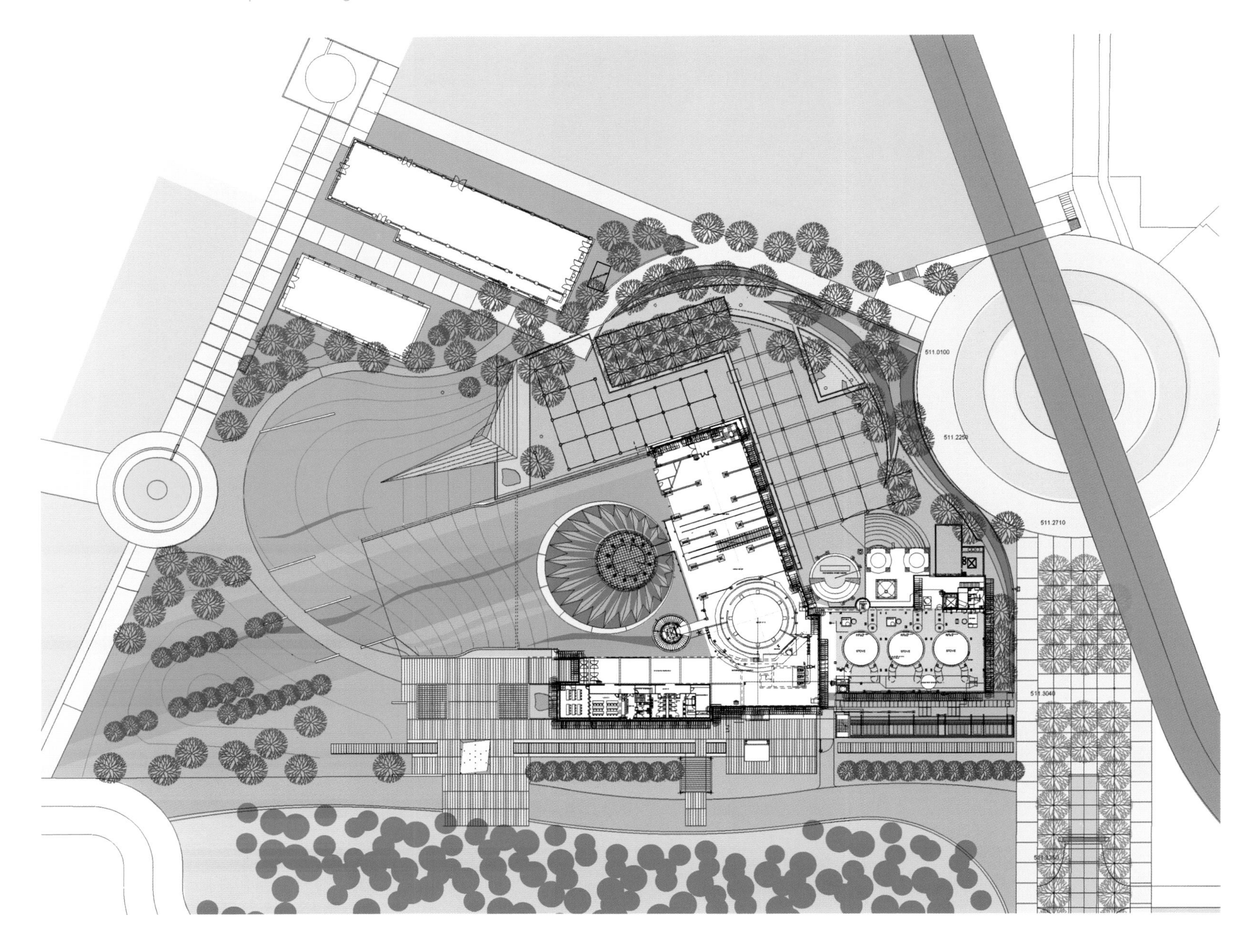

Museo Del Acero Horno³ Plan

总体的景观设计凸出了 70 m 高的鼓风炉的结构轮廓并和新建筑的现代设计相得益彰。大型不规则的钢铁和挖掘出来的机械装置合并成为步石和其他景观。这个设计方案通过使用绿色技术，将工业区开发、当地材料的适应性再利用与生态恢复融合起来。

两个水景组成叙事性设计，定义并定位博物馆旁边的公共空间。在主游憩区，一些以前用于主厅表面镀钢的钢材被重新用于建造有瀑布流过的台阶式水道。这个 200 m 长的水景暗示着每天运输成千上万吨原料并且在这里卸货的火车的运行轨道，同时与远处景观中的雨水花园形成视觉联系。在博物馆入口处，这条台阶式水道的尽头是水雾缭绕的喷泉和明显嵌入了矿石的网状岩石。这种错视画会引发腐蚀性加热，曾经被用来萃取矿石，但这种加热过程不会产生蒸汽，只会生成凉爽的吹拂在广场上的水雾，从而为身处蒙特雷干热气候中的游客带来惊喜。

博物馆上绿色屋顶（宽广集约型）的设计是拉丁美洲最大的屋顶体系，利于减少建筑物的视觉冲击。以前就有的鼓风炉结构位于这个屋顶之上。屋顶更高处则根据新建筑的结构性屋顶模式布置了多种抗旱性景天属植物，它们被限制在类似漂浮着的钢制磁盘的结构中。圆观景台供游客领略周围广阔的地域性景观，包括远处的马恩雷山脉，与屋顶土堆的外形遥相呼应。下面是佛得角地毯（绿毯），是一块不太规则的长着高茎草的草地——本地景观的一种抽象表现——与该景观曾经作为工业区的背景联系起来，既为退化的土地提供生物修复又增加新建筑的热工性。

阿塞罗奥尔诺博物馆景观设计的核心是可持续性原则。

# Civic Space Park

## 菲尼克斯文娱公园

Landscape Architect: AECOM

Photography: AECOM Photos by David Lloyd

AECOM's Design + Planning studio in Phoenix guided the project from visioning sessions and masterplanning of the park through complete construction observation. One of the park's standout features is a field of white columns beneath an undulating canopy that comes alive at night with light and color from an array of LED animations. Inspired by lightning touching down in an Arizona summer monsoon storm, the lighting show responds to the movements of visitors. An interactive water feature is another favorite activity in the park, especially for children.

The design of the park is low-carbon. Porous concrete paving and landscape design provide for stormwater collection and filtration, allowing every drop of water that lands on the site to recharge surrounding groundwater. Solar panels on the top of the park's shade structures generate 75 Kw of power, enough to power 8-9 residential homes and offset the park's lighting and electrical needs. When its trees and vegetation reach maturity, more than seventy percent of the park will be shaded from the desert sun. Most of these trees are deciduous in order to take advantage of the Arizona sunshine and mild climate in the winter months.

菲尼克斯市的 AECOM 设计公司通过完整的施工监理对公园的展望和总体规划进行指导。公园中一个比较显著的特色是波浪起伏的天蓬下面立着的数根柱子，夜间会因为一连串的灯管产生彩色的光而变得栩栩如生。受到亚利桑那州夏季季风风暴的启发，灯光的设计呼应了游客的活动。一个交互式的水景在公园里同样备受青睐，尤其受小孩子喜爱。

公园的设计理念是低碳。使用具有良好渗透性的混凝土铺砌和利用景观设计提供雨水收集和灌溉，使这片区域里的每一滴水都能被循环利用来补充地下水资源。良好的遮蔽性建筑顶部的太阳能板可以产生 75 Kw 的能量，足以为 8 到 9 个家庭供电，还可以补充公园里照明及其他用电需求。当树木和植物枝叶繁茂，公园里超过百分之七十的区域将会被树荫覆盖，从而躲过沙漠烈日的暴晒。大部分树木属于落叶型的，这样设计是为了利用亚利桑那州的阳光条件和冬季的温和气候。

该公园也包括一个艺术品设施，由国际知名的艺术家 Janet Echelman 根据 Ralph Waldo Emerson 的名言命名为“她的秘密是耐心”，灵感源自亚利

The park also includes an art installation from internationally recognized artist Janet Echelman titled, "Her Secret is Patience," named after a phrase by Ralph Waldo Emerson and inspired by elements of Arizona nature.

The southwest corner of the park features turf landscape forms with pedestrian-scale retaining walls, games tables, benches, and densely spaced shade trees — all of which are inviting for leisurely uses. A second phase of the Civic Space Park will connect to Phoenix's historic post office building to the north, expanding this vibrant, green thread in a desert cityscape.

桑那州的自然因素。

西南角区域的特点是草地景观，与挡土墙、游戏桌、长椅和浓密的树荫连成一体，为休闲活动提供空间。公园设计的第二阶段将会与北部的邮局连接起来，以扩大沙漠都市里这个充满活力的绿色地带。

UNIVERSITY C

# Promenade Samuel de Champlain

## 萨缪尔·德·尚普兰长廊

Landscape Architect: Daoust Lestage Inc. & Williams Asselin Ackaoui & Option Aménagement

Client: Commission de la Capitale nationale du Québec

Photography: Marc Cramer

The four thematic gardens are inspired by the river moods — the Quay-of-Fogs, the Quay-of-Floods, the Quay-of-Men and the Quay-of-Winds. They follow the geometry of four imaginary quays standing on the waving land. They bring a sequence of experiences and atmospheres, from the boundless visual expanse of the river and the scale of the territory, to the tactile sensory experiences of the human scale. They also integrate the Champlain Boulevard crossing it in a subtle way and bringing together the two sections of the linear park.

Each of this singular dense landscape captures and magnifies the material and poetic qualities of local coastal environment. One celebrates the winds and the bird flocks through whirling, poetic light-weight wind sculptures. The second one plays on man's archetypal response to tame by framing the water and the nature. The third one captures the river's grey waves and ice-brake patterns into the garden's water walls and vivid springs, playing along the rich textures and geometries of Quebec granite. The last one veils deep cross-river views and its monolithic boulders in the ever-shifting mist. The sinuous white concrete pedestrian path and the grey linear cycle track cross.

Promenade Samuel de Champlain Master Plan

The thematic gardens give structure, coherency and rhythm to the linear promenade which is materialized by the grey cycle track and the white concrete pavement of the pedestrian path. This project strongly contributes to revitalize an important section of the St-Lawrence River's borders and brings important wellness and economics repercussions to the city and the surroundings. It also contributes to bring, at a big scale, a contemporary face to Quebec City.

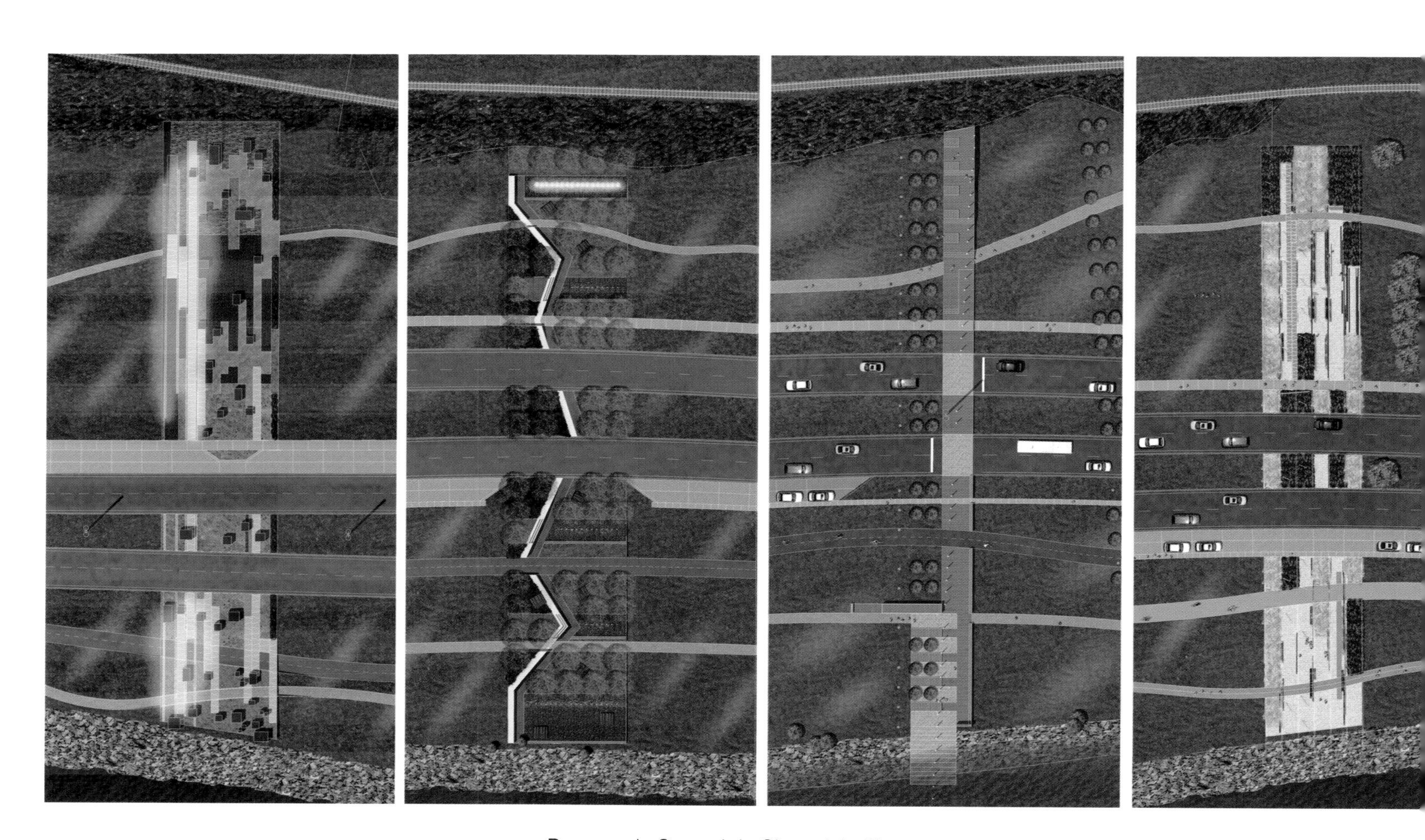

Promenade Samuel de Champlain Plan

四个主题花园是受到四个区域的启发：Quay-of-Fogs、Quay-of-Floods、Quay-of-Men 和 Quay-of-Winds。这四个码头是虚构的，伫立在绵延起伏的陆地上，花园的设计遵循了它们的几何结构。四个主题花园提供一系列的体验和氛围，从水域的宽广和土地的广袤，到人们自身的感官体验。另外，它们以一种微妙的方式将尚普兰大道融入并使之横穿于这个整体当中，同时将这个线形公园的两个部分结合起来。

每个密集的景观都捕捉到且放大了当地沿海环境的物理上的特性和诗意。第一个花园为风和鸟群设计，设有一些旋转着的诗意般的轻型风雕塑。第二个花园围绕水与大自然，依据着人对驯化行为的典型反应而设计。第三个利用了水域的灰色波浪、水墙和叮咚泉水中的破冰图案，展示了魁北克花岗岩多样化的纹理和几何结构。最后一个以其永恒变化着的雾霭模糊了渡河深处的视野及其巨大的岩石。横穿四个主题花园的是白色蜿蜒的混凝土步道和灰色的线型单车道，为线型的人行道提供了结构、凝聚力和韵律感，使之具体化。这个项目十分有利于复兴圣劳伦斯河边界的重要区域，为城市及周边环境带来积极的影响，促进经济的发展。同时它也极大地促进魁北克市现代面貌的产生。

# Foothill College

## 福德希尔学院

Landscape Architect: Meyer + Silberberg Landscape Architects

Client: Foothill DeAnza Community College District

Photography: Drew Kelly

Foothill College was originally designed in the early 1960's and has bestowed many top awards, including the ASLA Landmark Award. Over the past 50 years however, the campus has fallen into a state of disrepair. In 2008, a bond was secured to bring substantial infrastructural upgrades, focusing on a renewed sense of design and ecologically sustainability. The master planning efforts strive to bring vision to the landscape with the goal of invigorating the campus' social spaces while being respectful of the quiet dignity that permeates.

BAY

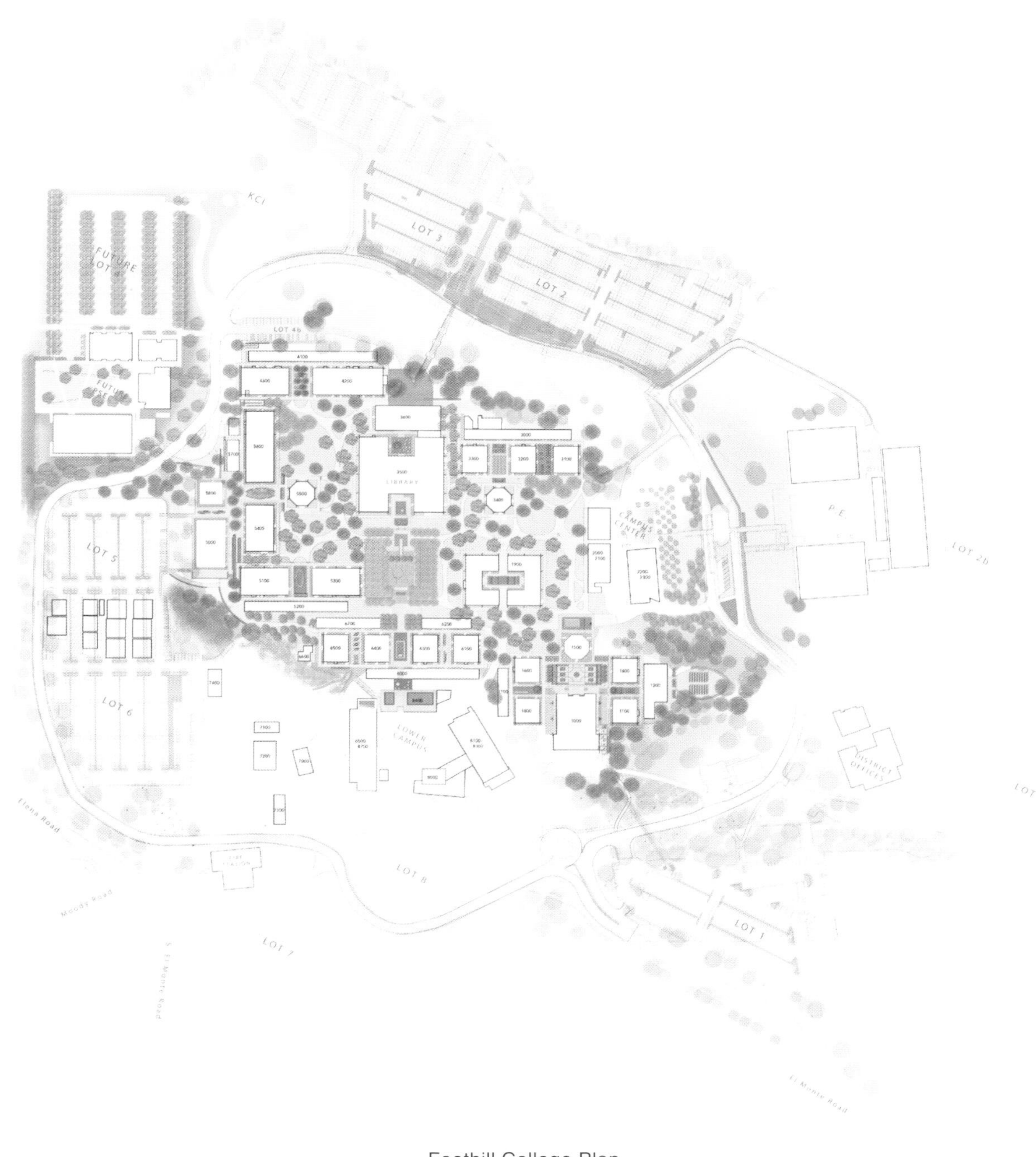


Foothill College Plan

福德希尔学院最初设计于 20 世纪 60 年代早期，曾被授予包括美国地标奖在内的很多高端奖项。然而，在过去 50 年里，这所大学已经沦落到一种失修的状态。在 2008 年，有一带被围起来以升级大量基础设施，项目注重设计理念的更新以及生态的可持续性。总体规划力图打造新的景观，最终为校园的社交空间注入新的活力，同时尊重空间中弥漫的安宁威严的氛围。

IDEA LAB
1200

# Mount-Royal Park's Playground

## 皇家山公园游乐场

Landscape Architect: Groupe IBI-CHBA

Client: Ville de Montréal

Location: Canada

Given this heritage designation, a dozen municipal and provincial organisations had to ratify this project, which included: a play ground conceived with a theme derived from Mount Royal itself, a picnic area in a grassy plain with approximately 30 tables, the redevelopment of roadways and paths. Cardinal Hardy was given the mandate to conceive the playground.

The theme is the Blue Spotted Salamander, an amphibian native to Mount Royal and the starring feature which organises the play structures and other park elements. Water features and other innovative play structures are integrated into the silhouette of the salamander as it rises from the earth; this instigates a different kind of play, which encourages the children's motor, cognitive and social development. Beyond simply contending with a heritage site, the project highlights the therapeutic influence of this large scale green space in the city.

The design was based on two distinct projections of the space; a vertically nuanced integration into the surrounding environment, and a horizontal plain contrasting the natural surroundings with the silhouette of the salamander and its bright colours.

Against this unusual backdrop, the landscape architect designed a Children Rights promenade of didactic elements. Public interpretive panels allow people of all ages to discover the rights guaranteed to children by the International Convention of Children's Rights.

考虑到这个遗产地，十几个市级省级机构批准了这个项目。它包括一个以皇家山为主题的游乐广场、一个草原上的设有30个桌子的野餐区和一些新开发的小路和车道。Cardinal Hardy受命设计这个游乐场。

设计主题是“蓝色斑点蝾螈”，一种本土两栖动物。这个独特的设计主题体现在游乐设施的结构以及公园的其他部分。水景和其他创新性的游乐设施被集中到蝾螈的轮廓中，仿佛它正从地面上爬起来。这种组合开发出很多的游戏机会，促进了孩子们的运动能力、认知能力和社交能力的发展。该项目突出了这片大规模绿色空间的健康与休闲作用，而不是一个简单的遗产地。

这个设计基于两个独特的空间规划：一个略带偏差的垂直建筑群和一块水平区域，以其蝾螈轮廓和明亮的颜色与自然环境形成对比。

景观设计师以此为背景设计了一个极具教育意义的“儿童权利”长廊。一些公共标语牌能让各个年龄层的人们了解国际儿童权利公约所认可的儿童权利。

être protégé

jouer

# North West Park

## 西北公园

Landscape Architect: SLA

Client: Municipality of Copenhagen

Location: Copenhagen, Denmark

Area: 35,000 $m^2$

Under the title "1001 trees" the park consists of four simple, but effective elements: trees, paths, light, and cone-shaped mounts. These elements create order and coherence between the park's many different parts. All four elements are distinct features of the park. Their simple, but varied compilation create a sequel of changing spaces and corners with altering atmospheres and feelings, which differentiate the park from the area's grey and fragmented environment.

The shapes and colors of the trees are completed by a "magic forest" of lamp posts and artificial lights. At daytime, one wanders through the architectonic composition of elegant lamp posts striped in different colors, cold ones in the north, warm ones in the south. At night, one is surprised by projected light, designing different patterns and colors on the ground.

North West as an area is going through changes. With SLA's design the park is a symbol and leader in this positive change. The North West Park is providing a run-down neighbourhood with an open park that protects and mirrors the diversity and changeability of the area. A park that embraces the need for everyone to feel welcome, while in quality competes with the finest urban spaces of Europe.

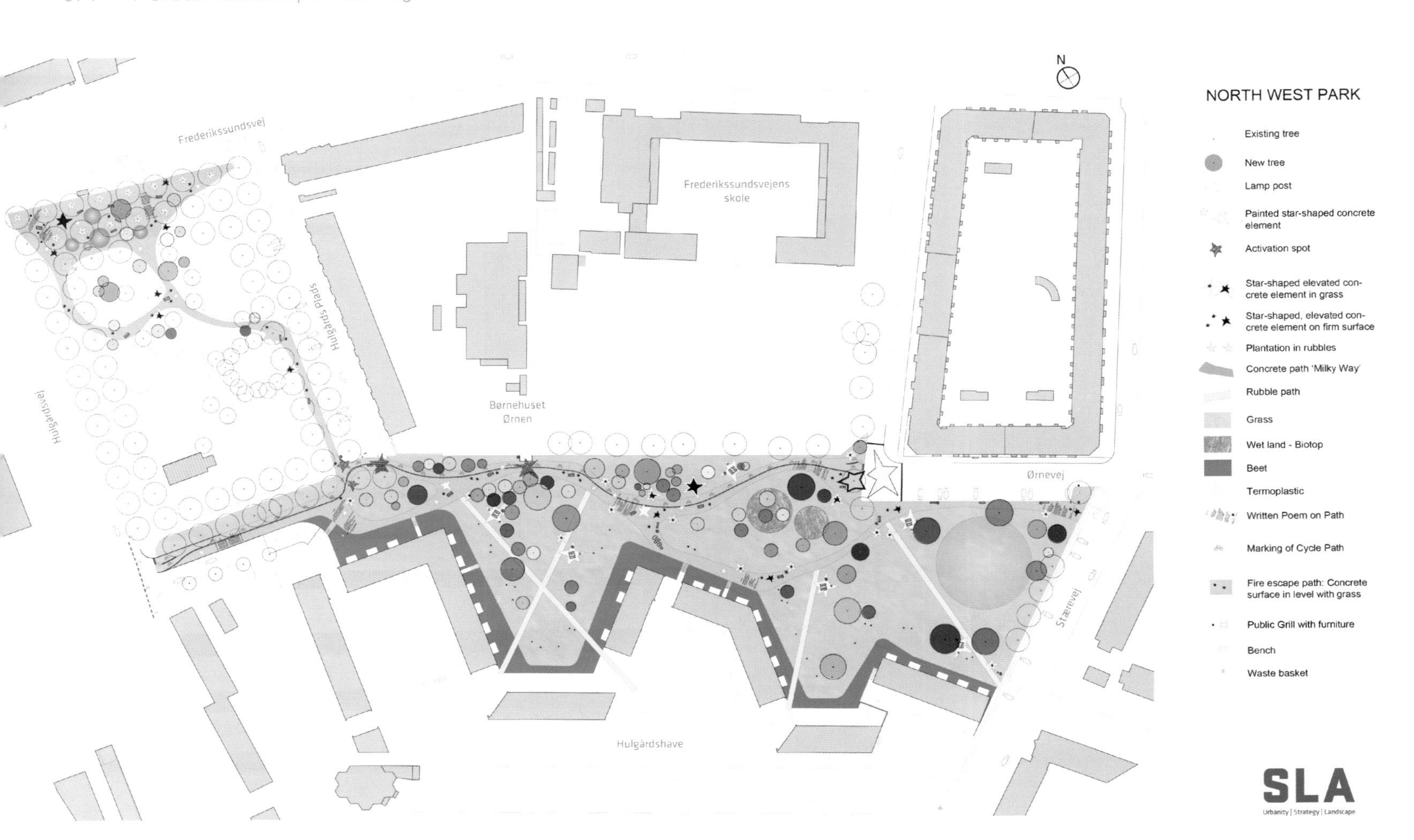


North West Park Master Plan

这个被称为“一千零一棵树”的公园由四个简单却重要的部分组成：树林、道路、光和锥形山体。它们使得公园里许多不同的部分有序而连贯地组合起来。这四个部分是公园的特色所在。它们之间简单而多样化的布局产生了连续变化着的空间和角落，同时改变着环境的氛围和人的感触，使公园从当地支离破碎的灰色环境中抽离。

由灯柱和人造光组成的“魔法森林”形成了树木的形状与颜色。在白天，当一个人漫游在雅致的带有彩色条纹的灯柱建筑组合中时，北面可以感受冷色调，南面可以感受暖色调。在晚上，人们会对投射光感到惊奇，那些光照在地上形成了不同的图案和颜色。

西北区正在经历着改变，而 SLA 的设计使得公园在这次积极的变迁中成为标志性的领导者。这个公园为周围荒芜的地区提供了一个既能保护又能反映该地区多样性和可变性的开放式公园。它让每一位游客宾至如归，同时在质量方面力图与欧洲最好的城市空间竞争。

# New Park for the University Quarter

## ▶ 大学校区公园

Landscape Architect: scape Landschaftsarchitekten

Water Management: Dahlem - Beratende Ingenieure GmbH & Co, Essen

Client: Grün und Gruga Essen

Location: Essen, Germany

Area: 5,1 ha

Photography: scape Landschaftsarchitekten, WAZ FotoPool Walter Buchholz

Paths, street furniture, lawn areas, water basins are designed with round edges like in foundry technology. All elements in the park — from plan view to detail — follow the same design and create the image of a unique park.

The surrounding residential and commercial buildings form an urban figure that gives the park a very long, narrow and radial shape. One of the most important design statements is made by arrangement of attractions and path relations, because of the particularly shaped park area, the request for water basins and varied recreational activities, as well as the expected high usage frequency of the inner-city park. To reduce potential conflicts between residents and park users secondary paths along the park edges are offered to strengthen the semi-public character. The main promenade is located in the spacious centre of the park along with recreation, sport and play facilities. The advantage is the possibility of a central, open, public promenade between the northern water park and the southern lawn park, as well as the reduction of noise for the residents.

Respectively to decreasing noise levels from Berlin Place to Viehofer Square, park facilities are assigned to subspaces. The entry point at Berlin Place is defined by a city square. Between the tree grove and the main event area, the section of the park intended for intensive use is situated, with its play and sunbathing areas and the "Promenade Place" called "Park Play". The subsequent section of the park at TurmstraBe is used for quiet recreation with its lawn areas and water basins as well as the "Promenade Place" called "Garden Park" .

New Park for The University Quarter Site Plan

小路、街道设施、草坪区和水池都设计了圆形边缘，像用铸造技术制成。公园的所有因素，从规划到细节，都是同一种设计风格，创造出一个独特的公园形象。

周围的住宅区和商业建筑物构成了一个城市形体，规划出公园狭长的放射状轮廓。对引人入胜的设施和路线的合理安排是最重要的设计之一。因为公园轮廓比较特别，需要设置水池举办多样化娱乐活动。城中公园高频率的使用需求，为了减少居民与公园使用者的冲突，公园边缘开通了第二条线路以巩固其半开放的特征。主干道位于宽阔的公园中心区，这里有提供消遣、运动及游戏的便利设施。这个设计的优势是在北边水公园与南边草坪公园之间提供一个集中的、开放的公共漫步道，以及能为居民降低噪音。

为了分别降低从 Berlin Place 宫到 Viehofer Square 广场的噪声等级，一些公园设施被分配到子空间里。一个城市广场规划了 Berlin Place 宫的入口。树林与主活动区之间坐落着公园的密集使用区，包括游戏区、日光浴区和被称为“Park Play”的散步广场。TurmstraBe 处公园的延伸部分包括草坪区、水池和名为“花园公园”的散步广场，为游客提供一个安宁的休憩区。

# Sa Riera Park

## 萨里埃拉公园

Design Team: Pere Joan Ravetllat, Carme Ribas, Manuel Ribas Piera, Carles Casamor, Marta Gabàs, Anna Ribas

Location: Palma de Mallorca, Illes Balears, Spain

Client: Town hall of Palma de Mallorca

Area: 290,000 $m^2$

Photography: Mandarina Creativos

The transformation of the wash bed and its surroundings to a city park means a significant increase in the green area of the city. It may still grow in the future with the incorporation of new areas currently with obsolete uses, so it would become the largest park in the city of Palma.

The park emerges as a big green area with soft surfaces and the capacity to reconcile its own uses with the specific city contingencies, such as parties, concerts and fairs. Located in the middle of the city, it means an interruption in the continuity of the urban fabric, not only because of its size but also for the depressed topography typical of a wash. To enhance the quality of the green reserves, the road traffic flows to the perimeter of the park and a new underground parking has been built under one of the side streets.

Rejected the possibility of covering the bed of the wash, The designers took advantage of its presence to structure the park in its longitudinal dimension. Section gradually widens generating a gradual approximation sequence to the dry wash bed.

The project is conceived from the section: it runs from the water course to the streets that confine the enclosure, building, the transition between city and park, and approaching the levels of the park to those of the current city. The asymmetric section of the wash bed opens to the upper esplanade of the park with green slopes.

The construction of the park is arranged from the manipulation of natural and rural images which are introduced in the city, taking advantage of that opposition: natural versus artificial. A geometric Mediterranean forest is located in the upper levels, understood as a viewpoint on the rest of the park. The intermediate areas are dealing with terraces planted as an agricultural crop. The lower esplanade is designed for more specific programs such as playground areas, bars or a monumental fountain.

这里的河床和周边被改造成城市公园，大幅度地增加了城市绿化面积。将来它可能合并一些废弃的土地，面积继续增长，成为帕尔马市最大的公园。

公园有着大面积绿地和柔和的外观，可为市里的聚会、音乐会和展览提供活动场所。坐落于城市中心，它的大小和典型的压缩性地形特征均使得城市肌理的连续性被打破。为了提高绿地的质量，公园由周边道路以及小巷地下新建的停车场来疏导周围的交通。

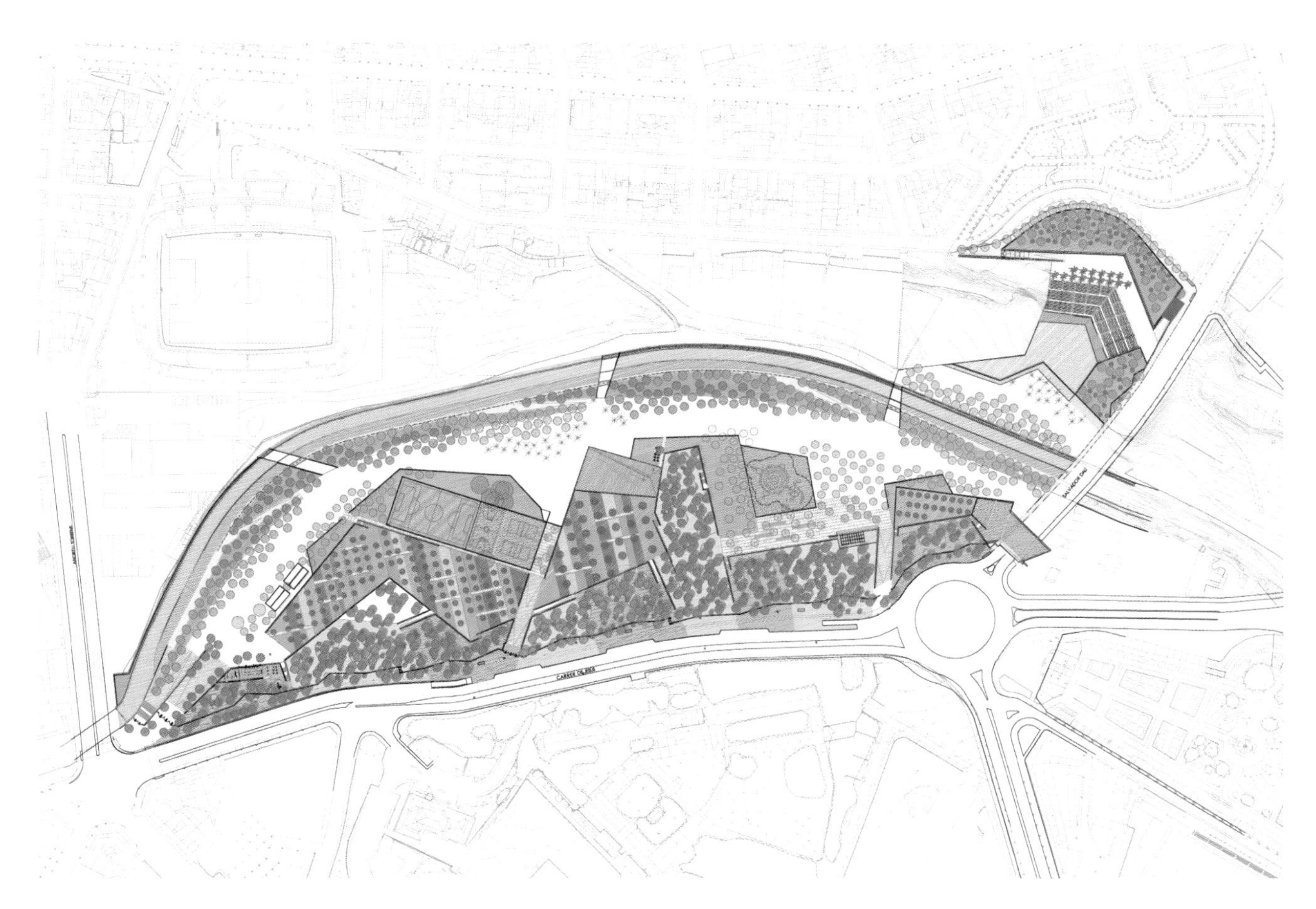

Sa Riera Park General Plan

设计无法完全覆盖河床，便利用它来构建公园纵向的景色。公园纵向的区域越大，便越接近干涸的河床。

这部分项目的构思如下：它将水从河道引入街道，和城市现有的其他公园一样，形成城市和公园之间的过渡。河床不对称的部分对公园绿色斜坡上的散步路开放。

公园建设从向城市引入自然和乡村景色开始，利用了自然与人工环境的差异。几何形状的地中海林区在公园较高的区域可作为观景点，俯瞰公园其余区域。中间的区域则种植了大量草坪。低处的平坦空地则为兴建游乐场、酒吧或纪念性喷泉等具体项目而保留。

# Parkstadt Schwabing

## Parkstadt Schwabing 商业园

Landscape Architect: Rainer Schmidt Landscape Architects + City Planners

Client: State Capital City Munich

Site Area: 400,000 m²

Photography: Raffaella Sirtoli, Michael Heinrich

Parkstadt Schwabing Plan

Parkstadt Schwabing商业园位于慕尼黑内城北部，是市区近期的开发项目，占地400 000 m$^2$。传统意义上慕尼黑的建筑设计偏向于朝南，面向阿尔卑斯山脉，便于一览其全貌。此种观赏方式可以从更高处观赏阿尔卑斯山脉，让其在众多主题公园中更显得别具生辉，这种设计也将成为市中心景区公园设计的主流手法，根据不同的公园和季节通过抽象表现法为游客营造不同的氛围。主题公园会让人们浮想联翩，如山卵石、高山湖、草地和公园的一处丛林。公园的最南端高达126 m的、突出的高塔是Parkstadt Schwabing项目的最大亮点。Parkstadt Schwabing项目空间结构的主要特点是它跟民宅和绿化带交相辉映。路过此地可以尽情一览此处的美景和风貌。大草坪和本地的桦树、松树和橡树又是公园美景的又一延伸。

中心公园长达700 m，宽达70 m，位于两排办公大楼之间。公园东侧专门设计成林荫大道和可散步的地方。而四分之一的街道还是步行甬道，强化了公园设计的节奏感。十字路口处的10 m高的精致楼阁缓解了高耸的办公楼和其他人工建筑带来的较压抑的视觉效果。

Parkstadt Schwabing is a 400,000 m$^2$ urban development project in the north of Munich's inner city. Munich traditionally prefers to look south, towards the Alpine foothills and the Alpine panorama. This viewing direction offering a view of the Alps from taller buildings, was to become the leitmotif of the local central landscape park, reflecting the different Alpine landscapes in themed gardens. These were conceived as an abstract representation of local landscape types revealing different atmospheres to everyday park users throughout the park and throughout the seasons. For instance, the themed gardens recall mountain boulders, an alpine lake, "meadows", and a forest within the single park. At the southern end of the park, the 126 m tall Highlight Towers crown the Parkstadt scheme. The main characteristics of Parkstadt's spatial structure are the staggered residential houses and the green streets. Pedestrians enjoy a wide range of views and perspectives. Large lawns and native trees such as birches, pines and oaks provide the experience of a continuous park landscape.

The central park, 700 m long and 70 m wide, is laid out between two rows of office blocks. The eastern flank of the park ribbon is earmarked for a boulevard-cum-promenade. The quarter's streets continue as walking routes in the park, giving it a rhythmic structure. The intersections are marked by delicate cube-shaped pavilions, 10 m high, mediating between the tall office buildings and the human scale.

# Punggol Promenade

## ▶ 榜鹅滨水步道

Landscape Architect: LOOK ARCHITECTS Pte Ltd.

Client: Urban Redevelopment Authority, Singapore

Location: Singapore

Photography: LOOK ARCHITECTS Pte Ltd.

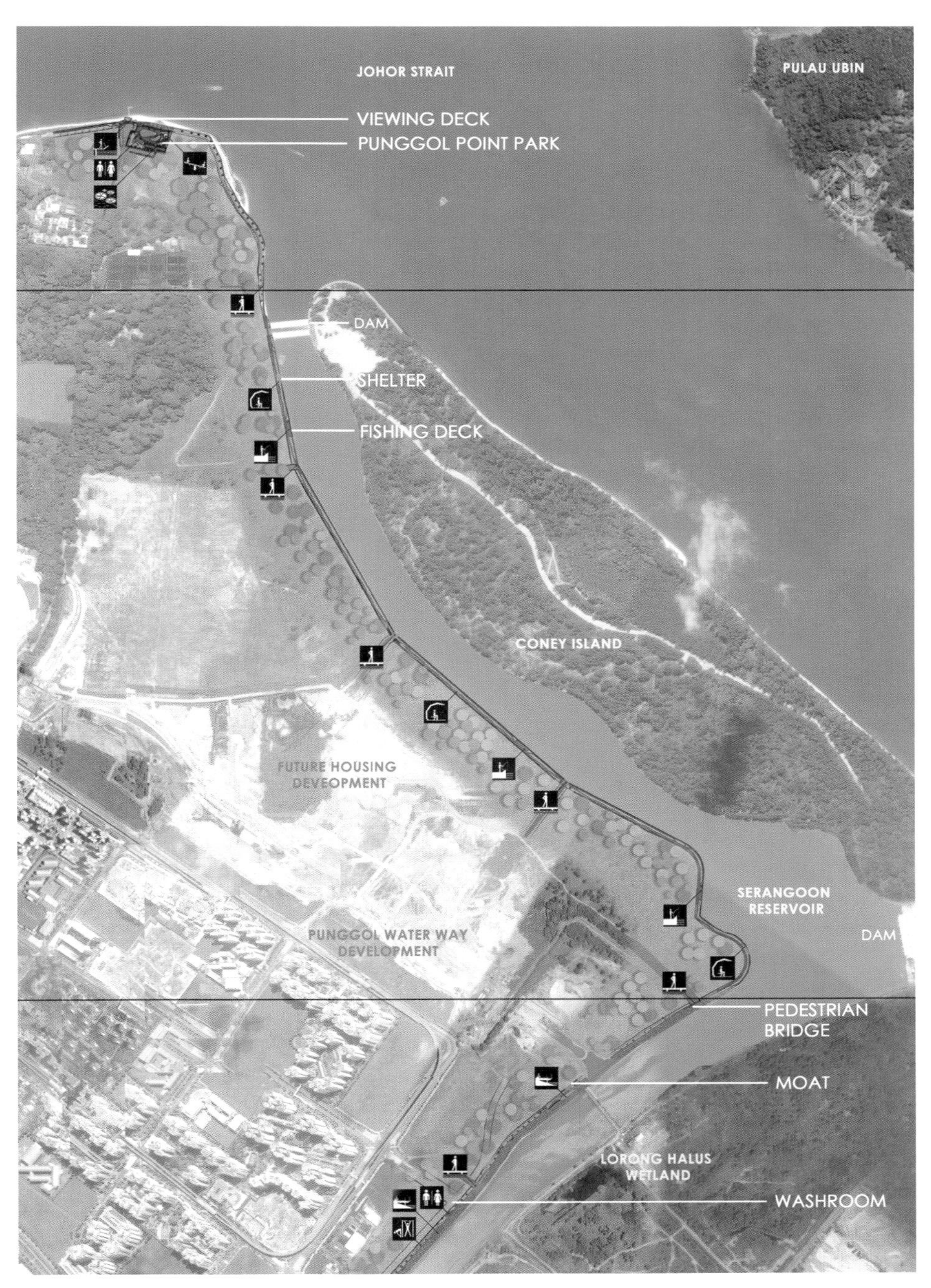


榜鹅长廊滨水景观成为了城市里的一首充满诗意的插曲，游客会在这时间和空间交错的环境里欢呼雀跃。它在保留榜鹅昔日风情的同时，也给人耳目一新的感觉。

这将是一次身临其境的个人体验，游客将备受大自然的抚慰和鼓舞。

在榜鹅海滩上，能看到水和天似乎无限延伸，模糊了周围高耸的公共建筑的轮廓，黑色混凝土基座上的安全栏杆引导着人们向地平线看去。

Punggol Promenade as a waterfront landscape insertion within the surrounding urban setting allows visitors to revel in a poetic interlude intertwining shifting planes of time and space — memories of a bygone era are evoked through embodiment of signifiers and reality melds with processes of re-familiarization to yield new layers of meaning.

The experience becomes so immersive and personal that a comforting sense of reconciliation with nature is inspired.

The expansiveness of water and sky seen from Punggol Point beach seems to stretch infinitely, obliterating the nearby towering neighbourhood of public housing, as the linearity of the safety railing sitting on black pigmented concrete plinths coaxes the eyes to be led towards the horizon.

大胆的色彩处理非常冷峻，在向作为二战纪念遗址的榜鹅海滩致敬，否则这段历史将鲜为人知。

人们可以看到俯瞰着榜鹅海滩的观景台与附近水域内油轮的船体非常相似，但它锋利的轮廓实际上呼应的是战争的另一个伤口。

The severity of this bold color treatment carries undertones of sobriety in tribute to Punggol Point beach as a World War II memorial site, a sliver of history that is otherwise little known.

One may see the viewing deck overlooking Punggol Point beach bearing resemblance to the hull of oil tankers dotting the surrounding waters, but its piercingly sharp silhouette may echo the scathing wounds of war for another.

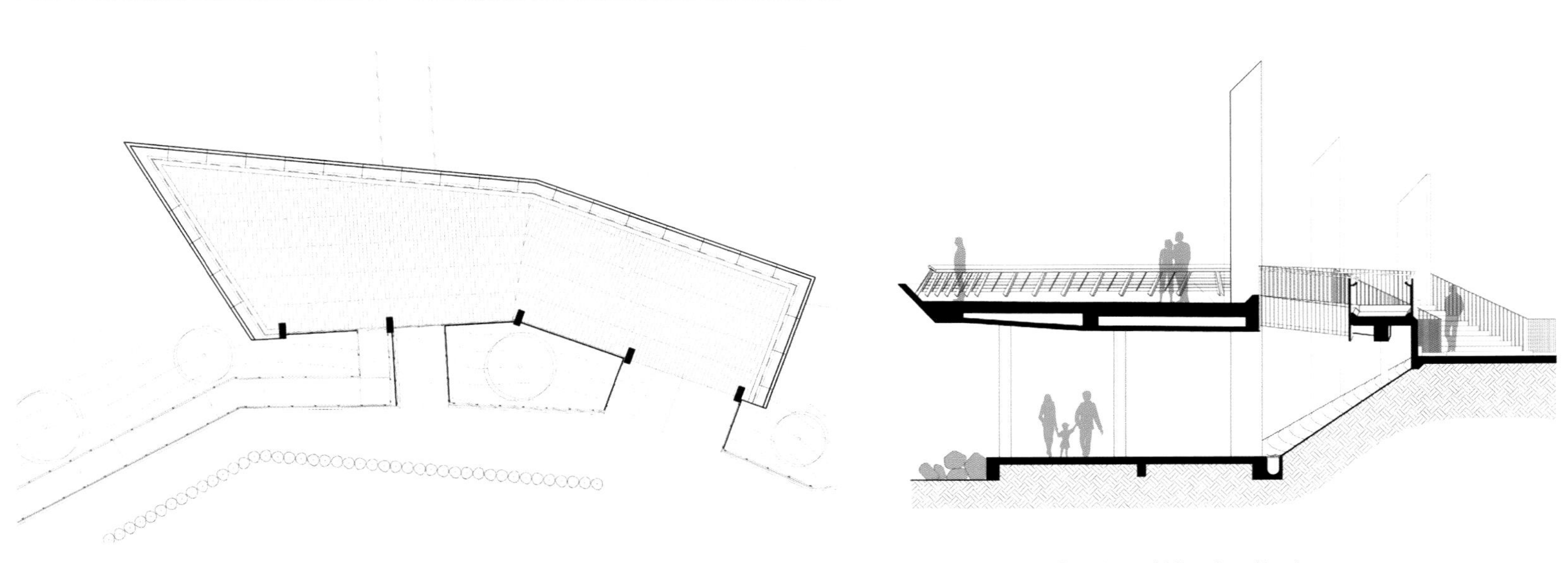

Plan of Viewing Deck

Section of Viewing Deck

Plan of Punggol Point Park

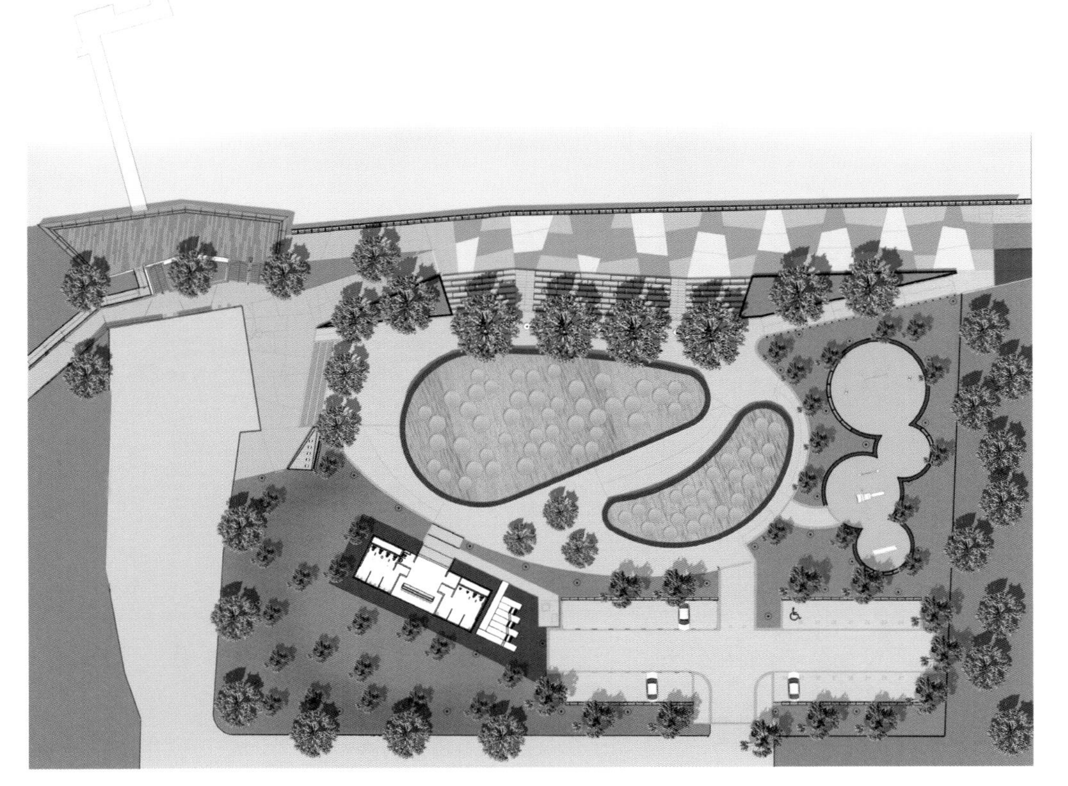

Plan of Punggol Point Park

Poetic speculation on man-made artifice and nature is inspired by calculated framing of tenderly stirred by calculated juxtapositions — whether glancing upon a lily pond by the seaside, or hovering close to the water edge on a cantilevered look-out platform — to rouse an invigorated connection to landscape and nature.

人工技艺和自然诗意的结合是设计师经过计算，设计并置而实现的，他们考虑了游客是否能一眼看到海边的荷花池，或能否漫步至悬臂式瞭望台旁边，感受景观和自然的鼓舞。

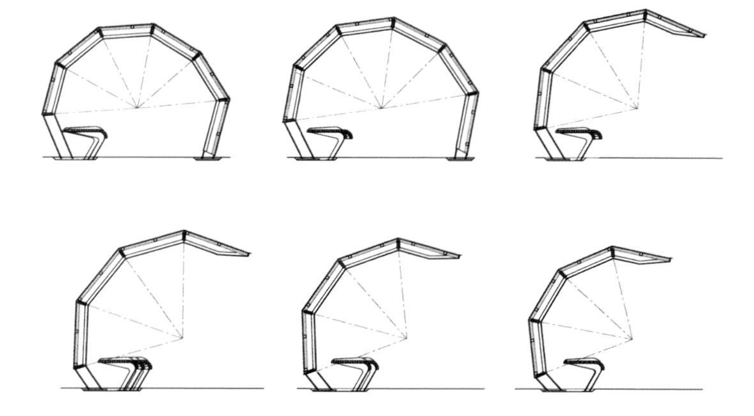

此项目以沿海的背景环境为灵感来源，模拟涌动的波浪以及螺旋状的贝壳的形态。

Shelter concept idea:
Inspired from elements of the coastal context, emulating the rolling sea waves and the corkscrew shell.

它铠甲般的外壳随着时间的推移将散发出风化的铜绿色光泽，结合了长廊的真实魅力。

内面则有着连锁的梯形铝板支撑外壳。休息亭散发出柔和的光泽，通过全天变化的光反射烘托周围的环境氛围。

The armor-like envelop would develop a weathered patina over time, the character of which melds with the down-to-earth charm of the promenade.

Inner faces of the shelter – interlocking trapezoidal aluminium panels strengthening the envelop – have a soft sheen which captures the ambient mood of surroundings by reflecting the changing hues of light throughout the day.

整个 4.9 km 长的海滨长廊所用的丰富的材料组合使人想起了旧时榜鹅的淳朴风情，它曾经是颇受部落欢迎的农场和种植园。

Composition of materials applied throughout the 4.9km long promenade make up a rich palette recalling the rustic character of old Punggol, a district which used to be populated with rural "kampong" communities keeping farms and plantations.

# Southport Broadwater Parklands

## 南港海滩公园

Landscape Architect: AECOM

Photography: AECOM Photos by David Lloyd

Southport Broadwater Parklands Site Plan

Ecologically sustainable development and water sensitive urban design features have been implemented as bold gestures integrated into the urban design of the parkland. They grab the attention of park users and present the notion of environmental responsibility and preservation as a tangible reality. Planted dunal mounding on the roof top of the pavilion building leads to an elevated viewing point. This initiative provides insulation to the building while challenging the traditional notion of a roof and showing what benefits can be achieved through an integrated design approach.

255 photovoltaic panels are integrated into shade shelters to provide a solar spine linking the city to the water. This consolidated display of panels offsets power requirements for stage 1 park lighting in an integrated, architecturally striking manner.

A 5-hectare catchment of previously untreated CBD stormwater runoff is captured and cleansed in a sculptural, terraced urban wetland before discharging to the Broadwater. This feature not only provides environmental benefits but also creates a quiet, contemplative "breathing space" between the two large-scale event lawns on either side.

生态可持续发展和水敏性城市设计作为海滩公园设计中显著的一部分已经开始实施。它们吸引了公园使用者的注意力，提出将环境责任和环境保护变成实实在在的现实。凉亭顶部的绿色土丘将人们的视野引向更高处。这种设计将建筑孤立起来，挑战了屋顶设计的传统观念，并且显示了完整的设计方案所产生的效果。

遮蔽设施中增加了 255 个光电面板以提供连接城市与水域的太阳能。 这样的设施以一种惊人的建筑方式满足了第一阶段公园内照明的需求。

在商务中心区有一个 5 公顷的之前没有被处理过的流域，一个雕塑般的台阶式湿地用来收集并净化雨水，然后排放到沙滩区。这个景观不仅带来良好的环境效益，而且在两边的大片活动草坪地之间创造出一个宁静、引人深思的“呼吸空间”。

The pairing of a clear civic or urban program within the park with the ever-present riparian condition defines a strong sense of the park's role as a place for major community gatherings. The main events lawn will play host to some of the city's largest and most exciting events, including the Gold Coast Airport Marathon; the ITU World Triathlon Championships, which attracts over 25,000 participants and spectators; Carols by Candlelight at Christmas; and numerous other community events throughout the year.

Public artwork is based on the theme of "Currents and Catchments", a broad metaphor to capture the sense of openness, movement and natural flows of activity that characterize the parklands. John Tonkin, an electronic artist commissioned through our art consultant, created a "nervous system" consisting of 20 light and sound-emitting poles.

一项清晰的公园中的城市规划与时时存在的水滨条件的结合，定义了公园作为主要的社区集会场所的重要角色。主要的活动草坪上将会举办一些城市里最大型、最激动人心的活动，包括：Gold Coast Airport Marathon 马拉松比赛；ITU World Triathlon Championships 冠军赛，这个比赛吸引了25,000 名参加者和观众；Candlelight 表演的圣诞颂歌以及全年众多的其他活动。

公共艺术品是以“潮流与汇集”的主题为基础，隐喻着捕捉海滩公园活动的开放性、运动性和自然流动性。电子艺术家 John Tonkin 应邀做艺术顾问，创造了一个包含 20 个发光发声电线杆的“神经系统”。

# 5 Star Hotel Ifen

## 伊芬五星级酒店

Landscape Architect: glasser and dagenbach, landscape architects bdla, IFLA

Client: Travel Charme group

Location: Hirschegg, Kleinwalsertal, Austrian alps

Photography: Udo Dagenbach

The Travel Charme group decided to restart and refurbish a very traditional 5 star Hotel in a small valley in the beautiful located Kleinwalsertal in the Austrian alps. The site is located on a steep ground with 14 m height difference on 1,100 m above sea level. It is a mountainous region with mountains up to 3,000 m and a very traditional summer and winter tourism area.

The architects' job was to design the 1.9 hectare area according to the regional ecological situation. A big part of the terrain is so steep that normal plantings and any use is not possible. To cut down costs they decided to install a grass landscape with mountain species which are adapted to the local conditions.

For the shape of the terrain is very inhomogeneous we had the idea to spread leaf shaped flowerbeds all over the site like thrown by chance or blown there by winds. The flowerbeds are cut into this paving. The shown pictures are taken two weeks after planting. Many local rocks are used for walls and stone settings. Main trees are Sorbus intermedia, Pinus nigra nigra and Acer neglectum annae. Shrubs like Sambucus nigra and many mountain roses are used.

The main idea of the concept: The best garden is the nature around – let it be the best – don't try to work against it – enhance it by little impact.

5 Star Hotel Ifen Site Plan

旅程魅力组决定重新开始整修一个五星级酒店，它位于奥地利阿尔卑斯山美丽的 Kleinwalsertal 区中的一个小山谷里面。酒店坐落在一个 14 m 高陡峭的地面上，海拔高度为 1 100 m。这是一个多山的地区，最高至 3 000 m，也是传统的夏季和冬季度假区。

建筑师的任务是根据当地生态现状来设计这片 1.9 公顷的区域。大部分地势十分陡峭，因此常规种植是几乎不可能的。为了降低成本，他们决定装置一个草地景观，用来种植当地的本土植物。

由于地形参差不齐，设计师打算在整个项目区域铺开叶形花圃，使之呈现出如同被偶然抛到地面或是被风吹开的样子。花圃被切分铺到阶地上。那些展示的图片是在种植的两周后拍摄的。一些当地的岩石被用来筑墙或布置石景。树木主要是楸树、黑松和槭树。这里也种植了西洋接骨木和野玫瑰等灌木。

设计概念的重点：大自然是最好的花园 – 尊重这个原理 – 不能违背大自然 – 以细微之举完善它。

# Tanner Springs Park

# 美国波特兰坦纳斯普林斯公园

Landscape Architect: Ramboll Studio Dreiseitl

Local Landscape Architect: GreenWorks PC

Client: City of Portland

Location: Portland, USA

Area: 4,000 $m^2$

Copyright: J Hoyer

Formerly a wetland, the Pearl District was bisected by Tanner Creek and sided by the broad Willamette River. Rail yards and industry first claimed and drained the land. Over the past 30 years, a new neighborhood has progressively established itself — young, mixed, urban and dynamic, today the Pearl District is home to families and businesses. With surgical artistry, the urban skin of one downtown block, 60 m x 60 m is peeled back to create a new city park. Stormwater runoff from the park block is fed into a natural water feature with a spring and natural cleansing system. The "Art Wall" recycles historic rail tracks, oscillating in and out and inlaid with fused glass pieces hand-painted with nature images by Herbert Dreiseitl. Ospreys dive into the water, art performances unfold on the floating deck, children splash and explore, and others take quiet contemplation in this natural refuge in the heart of the city. An intense community participation and a stakeholder steering group means that this park is the realization of the dreams and hopes of local people.

Copyright: Ramboll Studio Dreiseitl

Copyright: Tom

Copyright: Ramboll Studio Dreiseitl

波特兰珍珠区基址原为一片清泉滋润的湿地，被坦纳河 (Tanner Creek) 从中划分开来，与宽广的威拉麦狄河 (Willamette River) 相邻。铁路站和工业区首先占用了这片土地，并伴有场地排水要求。在过去的 30 年里，一个新的社区被逐步建成，它象征着年轻、综合、大都市和活力。今天的珍珠区已经成为了商业和居住区域。在一个市区繁华地带大约 60 m × 60 m 的地方，重新塑造一个崭新的城市公园。从公园街区收集的雨水汇入由喷泉和自然净化系统组成的天然水景。从铁路轨道回收的旧材料被重新利用并建造公园中的“艺术墙”，唤起人们对于历史铁路的记忆，而波浪形的外观设计则能够给人以强烈的冲击感。戴水道设计公司的创始人赫伯特·德赖塞特尔先生本人通过手绘，将这里曾经生存的生物图案绘制于热熔玻璃上，并镶嵌在“艺术墙”内。在这个繁华的市中心地带，生态系统得到了恢复，人们居然可以看到鱼鹰潜入水中捕鱼。在甲板舞台上可以尽情地表演各种文艺活动，孩子们来到这里玩耍、探索自然奥秘，而另外一些人们则可以在这片自然的优美秘境中充分享受大自然的芬芳、进行无限的冥想。深入的社区参与和地产调查显示，这个公园是当地人们实现梦想和希望的地方。

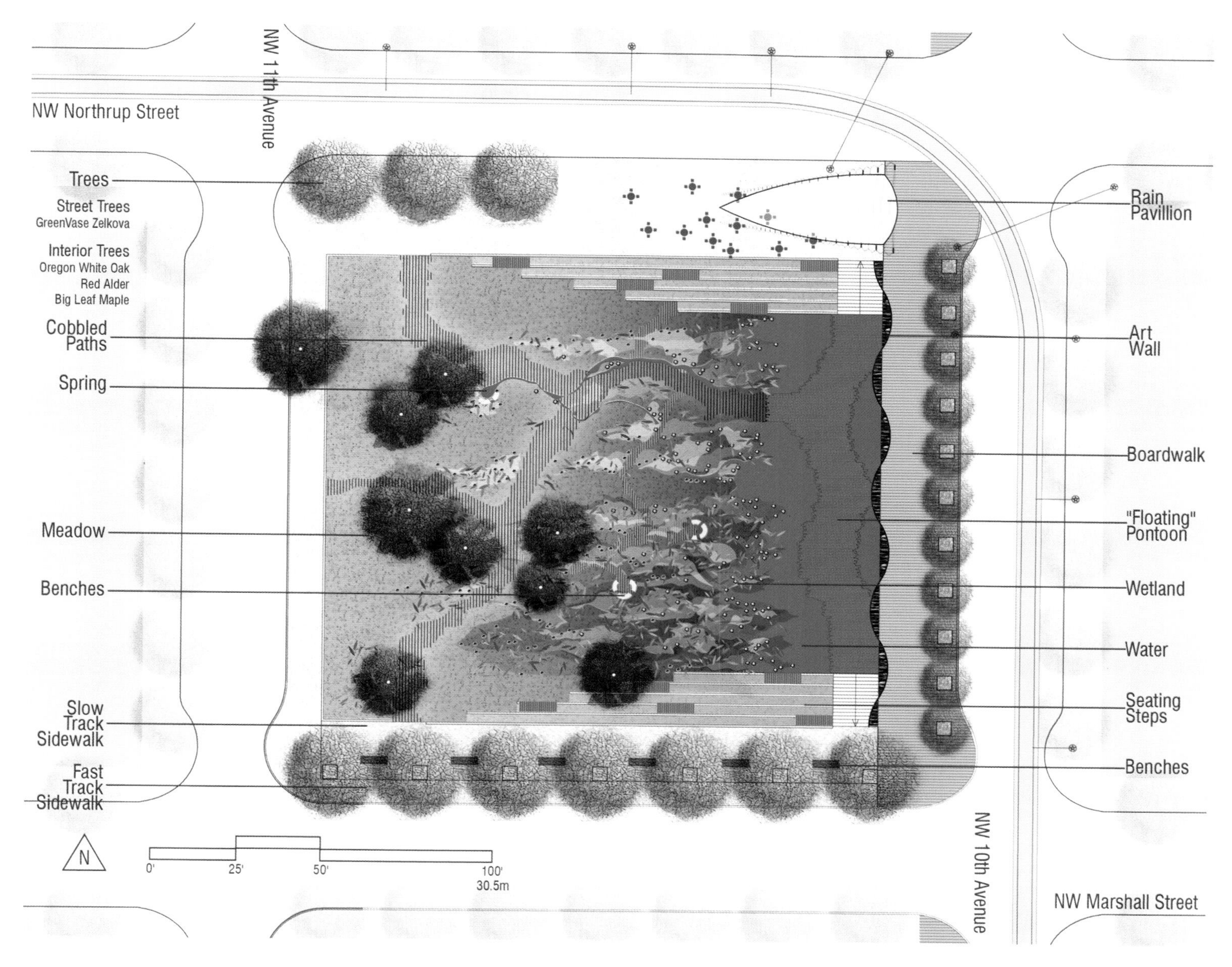


Tanner Springs Park Plan

Tanner Springs Park Section

Copyright: Ramboll Studio Dreiseitl

# The Crystal — A Sustainable Cities Initiative by Siemens

## ▶ “水晶”—— 西门子的可持续城市倡议

Landscape Architect: Townshend Landscape Architects

Client: Siemens plc (Planning), London Borough of Newham (Delivery)

Location: London Borough of Newham, United Kingdom

Area: 18,000 m$^2$

Townshend Landscape Architects were appointed by Siemens and the London Borough of Newham to ensure that high quality public spaces were designed and delivered, for the use and enjoyment of both visitors and the local community. The aim was to create a landscape that is distinctive yet maintains the area's character and identity, providing a richer, more engaging landscape.

The team worked closely with the London Borough of Newham and Design for London to create a framework for establishing a "Community Strategy" for the centre. This involved hosting regular events, workshops, gardening days and information on urban beekeeping.

The external spaces at the centre were designed to foster an awareness of the environment, inspiring people to think about their ecological footprint. In particular it highlights that food production has a significant environmental impact. It is estimated that food is responsible for 41 per cent of London's "ecological footprint". In response a series of gardens were created to promote cultivation in the community. The gardens use a variety of different planting concepts to showcase "sustainable" planting techniques as a tool for both education and research. The objective was to promote future urban environments to have a richer native vegetative base with outstanding wildlife quality. The design for the landscape sought to strike a balance between looking good yet providing a wildlife value with a high nectar source, for bees and other insects. Native wildflower meadows and traditional flower gardens demonstrate this potential inspiring people to think about their own back gardens, balconies or window boxes.

The scheme is the first project that has achieved "Outstanding" BREEAM and "Platinum" LEED.

The Crystal Illustrative Landscape Master Plan

西门子和伦敦纽汉区委任汤森景观建筑公司设计开发高质量的公共空间，供游客及当地社区享用。设计旨在创造一个既与众不同又保留当地特性和身份的内容丰富的迷人景观。

设计团队与伦敦纽汉区和伦敦设计公司密切合作，为建立一个市中心的“社区战略”提供方案，包括定期举办活动、开研讨会、园艺设计讨论和城市养蜂信息收集。

项目中心的外部空间设计致力于提高环境保护意识，鼓励人们反思他们的生态足迹，尤其强调了粮食生产对环境的显著作用。据估计粮食在伦敦“生态足迹”里占了 41%。因此，这里设计了一系列花园，以提高耕作能力。这些花园利用各种种植概念展示作为教育工具和研究工具的可持续种植技术。设计的目标是提高未来的城市环境质量，使之拥有丰富的本土植被和出色的野生环境。设计师为蜜蜂及其他昆虫设计了高级花蜜源，力图在好看与提供野生动植物价值之间取得平衡。本土的野花草地和传统花园彰显了激发人们思考自己的后花园、阳台或开窗盒的潜力。

这个设计是获得 BREEAM“优秀”和首个 LEED 白金级认证的项目。

# Shoemaker Green

## 口袋公园

Landscape Architect: Andropogon Associates Ltd.

Client: University of Pennsylvania

Location: University of Pennsylvania, USA

Area: 15,176 $m^2$

Photography: Barrett Doherty and Andropogon Associates Ltd.

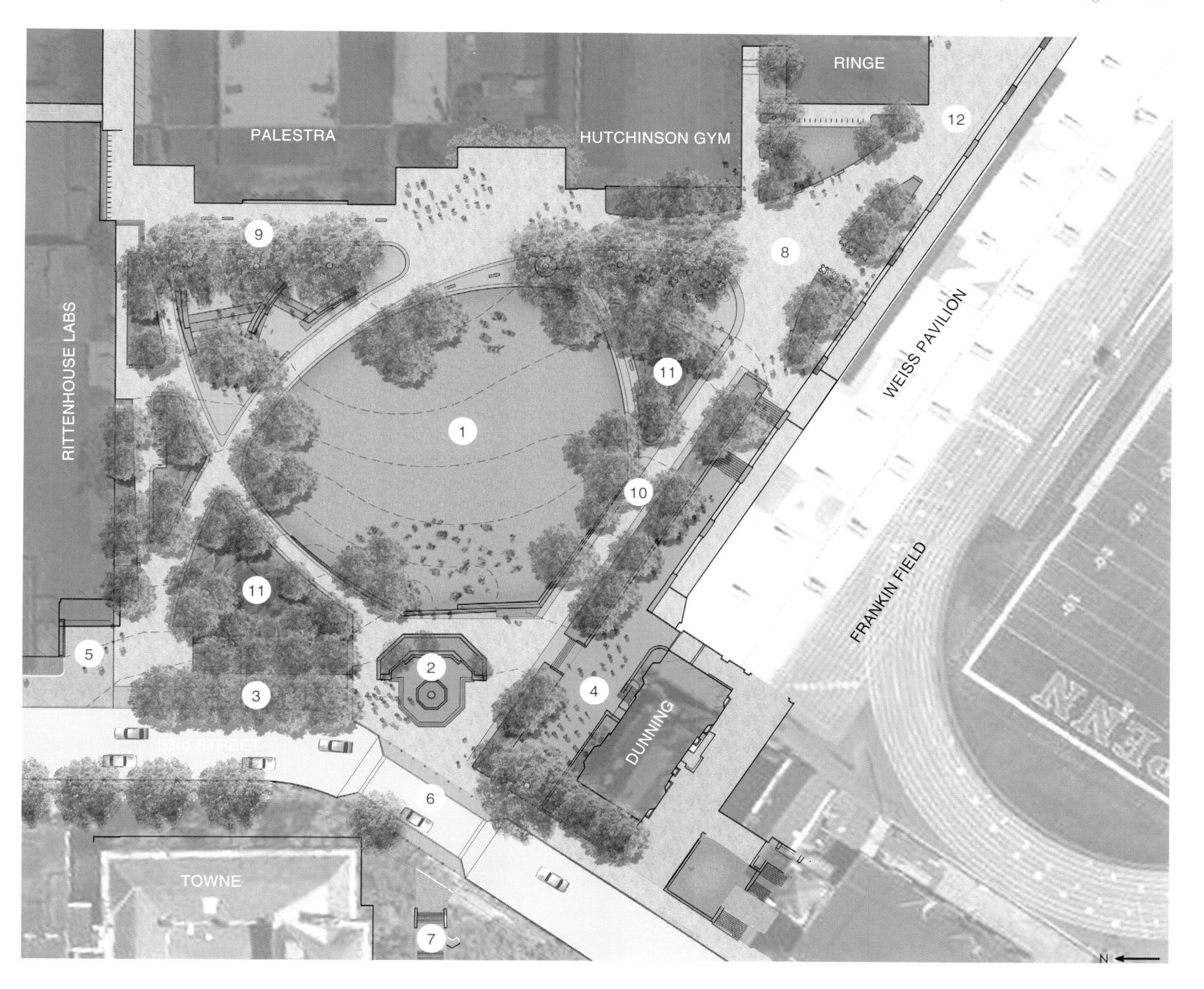


1. SHOEMAKER GREEN
2. WAR MEMORAIAL
3. 33rd STREET TREES
4. WEISS / DUNNING COURT
5. DAVID RITTENHOUSE ENTRANCE PLAZA
6. PEDESTRIAN CROSSING
7. SMITH WALK (EXISTING)
8. WEISS TERRACE
9. PALESTRA / HUTCHINSON TERRACE
10. SMITH WALK (PROPOSED EXTENSON)
11. RAIN GARDENS
12. CONNECTION TO PALEY BRIDGE / PENN PARK

Through the innovative use of a variety of strategies and technologies, the design of Shoemaker Green has been optimized to capture and control storm water from the site and surrounding rooftops, provide viable native plant and animal habitats, minimize transportation of materials to and from the site, and serve as a starting point for the development of a sustainable maintenance strategy for the University at large.

Rainwater management is provided by a landscape-integrated stormwater management system that includes the conveyance, capture, filtering, and storage of stormwater for reuse for irrigation purposes. The site was designed to capture and reuse 95% of the site's rainwater. Visitors are invited to engage with the rainwater management design where it is made visible in a rain garden. Stone weirs convey the water while plantings and gravel beds capture and filter the water. All the 43 different species planted are native to the Piedmont and Coastal Plain eco-regions and were procured locally, within 241.4 km of Philadelphia. Shoemaker Green is a model of sustainable campus design. Ongoing monitoring of the site will help inform future University landscape design and, as one of the Sustainable Sites Initiative's (SITES™) Pilot Projects, the green will help establish national sustainable landscape guidelines as well.

Water Diagram

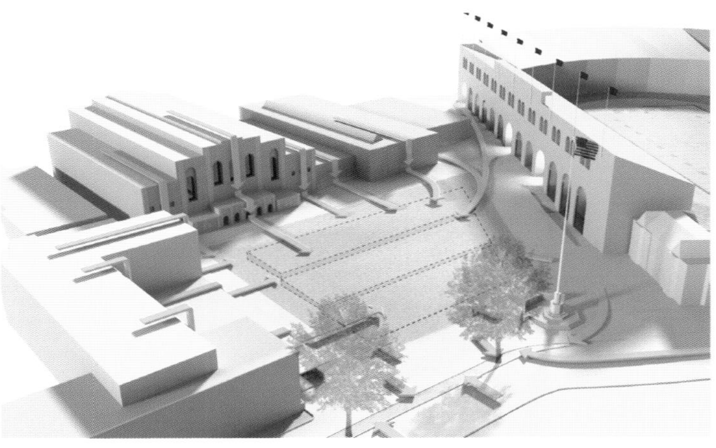

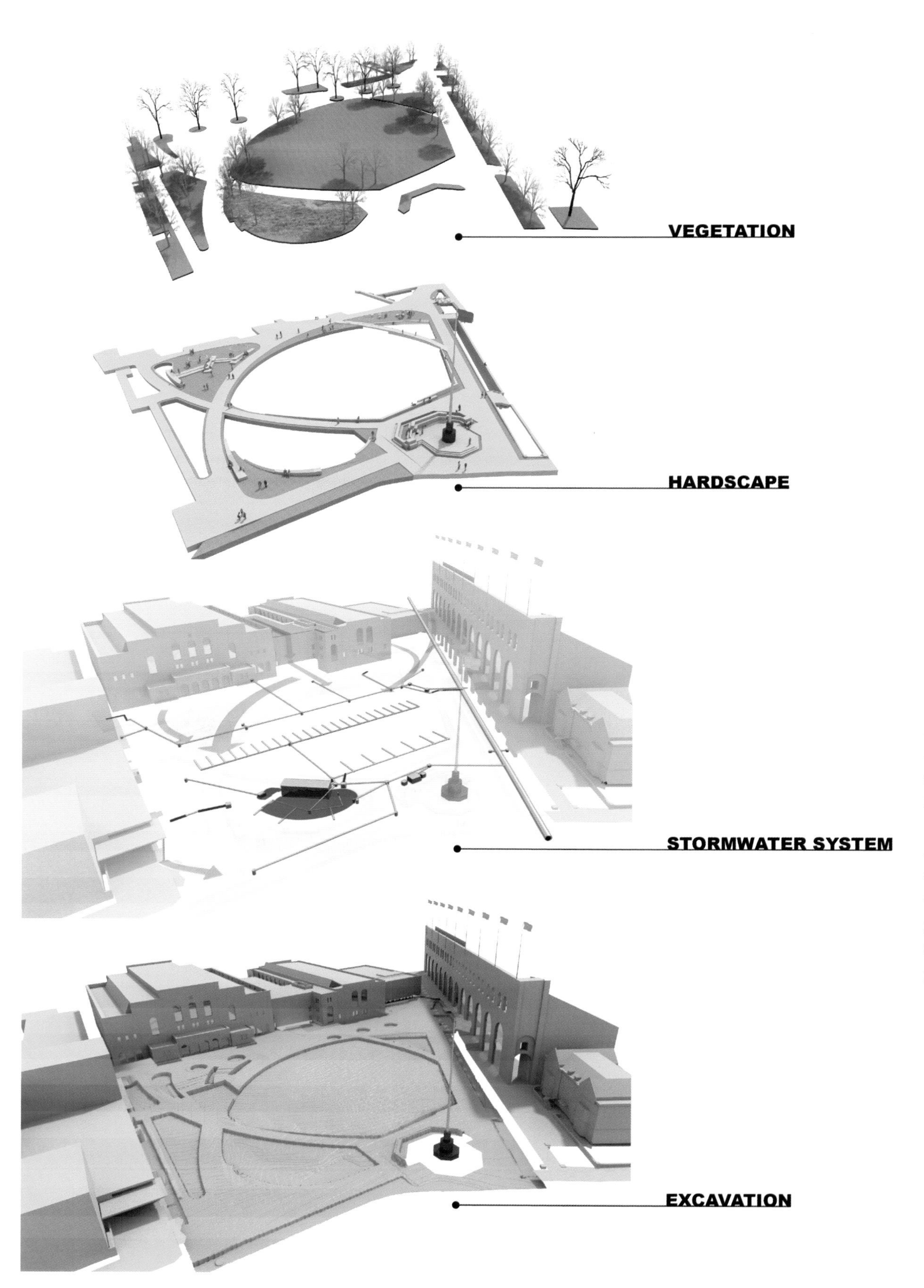


Exploded Axon

通过一系列的策略创新和技术创新，口袋公园的设计被优化到最佳水平，以收集并控制来自当地及周围屋顶的雨水，为本土植物和动物提供栖息地，使当地材料运输量减到最低。该项目也是大学发展可持续性维护战略的起点。

雨水管理由综合景观的雨水管理体系提供，该体系包括运输、收集、过滤和供灌溉再利用的雨水储存。该设计旨在收集并再利用当地95% 的雨水。雨水管理设计也考虑到游客的参与，在雨水花园中可以看得见。石坝能够运输雨水，植被和砾石层则收集并过滤雨水。43 种不同物种是 Piedmont 和 Coastal Plain 的生态区域的本土植物，产自费城 241.4 km 以内的地区。口袋公园是一种可持续性校园的典范。该地的持续监测有利于展示未来的大学景观设计，作为 SITES ™的试验项目，这个绿色空间也有助于建立国家可持续性景观的指导方针。

# Bishan-Ang Mo Kio Park and Kallang River

## 碧山宏茂桥公园和加冷河

Landscape Architect: Ramboll Studio Dreiseitl

Client: Public Utilities Board & National Parks Board

Location: Singapore

Area: 620,000 m$^2$

Copyright: Public Utilities Board & Ramboll Studio Dreiseitl

Bishan-Ang Mo Kio Park is one of Singapore's most popular parks in the heartlands of Singapore. As part of a much-needed park upgrade and plans to improve the capacity of the Kallang channel along the edge of the park, works were carried out simultaneously to transform the utilitarian concrete channel into a naturalised river, creating new spaces for the community to enjoy. At Bishan-Ang Mo Kio Park, a 2.7 km long straight concrete drainage channel has been restored into a sinuous, natural river 3.2 km long, that meanders through the park. 62 hectare of park space has been tastefully redesigned to accommodate the dynamic process of a river system which includes fluctuating water levels, while providing maximum benefit for park users. 3 playgrounds, restaurants, a new look out point constructed using the recycled walls of the old concrete channel, and plenty of open green spaces complement the natural wonder of an ecologically restored river in the heartlands of the city. This is a place to take your shoes off, and get closer to water and nature!

Copyright: Ramboll Studio Dreiseitl

Copyright: Ramboll Studio Dreiseitl

Copyright: Ramboll Studio Dreiseitl

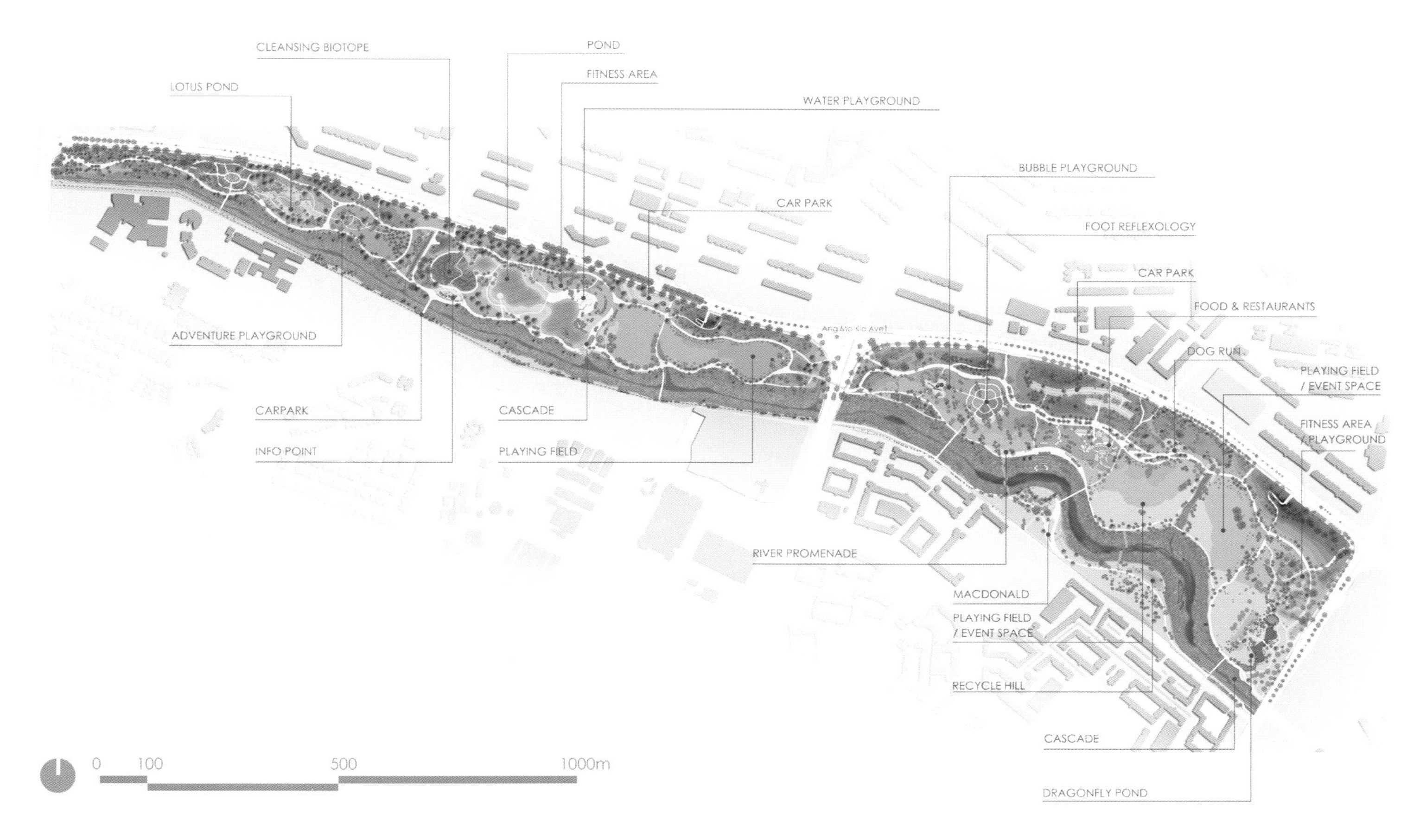


Bishan-Ang Mo Kio Park and Kallang River Plan

碧山宏茂桥公园作为新加坡市中心区域最受欢迎的公园之一，急需规划升级。与此同时进行的还有公园沿线的加冷河水渠修复计划，将单一功用性的混凝土结构河道转变为自然式河流，以创造社区居民能够充分享用的新型城市空间。在碧山宏茂桥公园，长 2.7 km 的笔直混凝土排水渠已经恢复为长 3.2 km 的弯曲、自然式河流，蜿蜒穿过公园。62 公顷的公园空间被重新进行设计，以适应包括水位波动等的河流系统的动态过程，为使用者提供最大化收益。使用原有混凝土水渠改造而来的回收石材，建造了 3 个游乐场，餐厅，以及全新的瞭望台。大量的绿色开放空间也为城市中心区生态修复河流形成的自然奇观提供了有益的补充。这里是一处可以脱掉鞋子，与水和自然亲近的地方！

# Shenzhen Bay Coastal Park Master Plan

## 深圳湾公园总体规划

Landscape Architect: SWA Group

Client: Shenzhen Bureau of Planning

Location: Shenzhen, China

The project area includes approximately 15 km, encompassing three distinctive coasts: the west coast, the bay waterfront and the east coast. The proposed master plan seeks to unify the entire coastline into a cohesive whole. The landscape architects' solution begins with the restoration of large areas of mangroves and the re-establishment of sea grasses and salt marsh vegetation.

The landscape architects defer to the experience and expertise of China's own engineers to reclaim land from the sea but also recommend procedures for protecting and enhancing the shoreline, such as heavy stone edges. The plan proposes the building of four hills along the water's edge to provide visual reference points as well as viewpoints to the bay and the city.

The master plan also proposes a coastal ecological education center which will be an education and exhibition showcase to teach students, local visitors, tourists and researchers about the physical, biological and social conditions of the water's edge. In addition to indoor facilities, the center will include an outdoor coastal aviary, demonstration tidal pools, educational boardwalks and teaching pavilions. The boardwalks will extend into the salt grass areas.

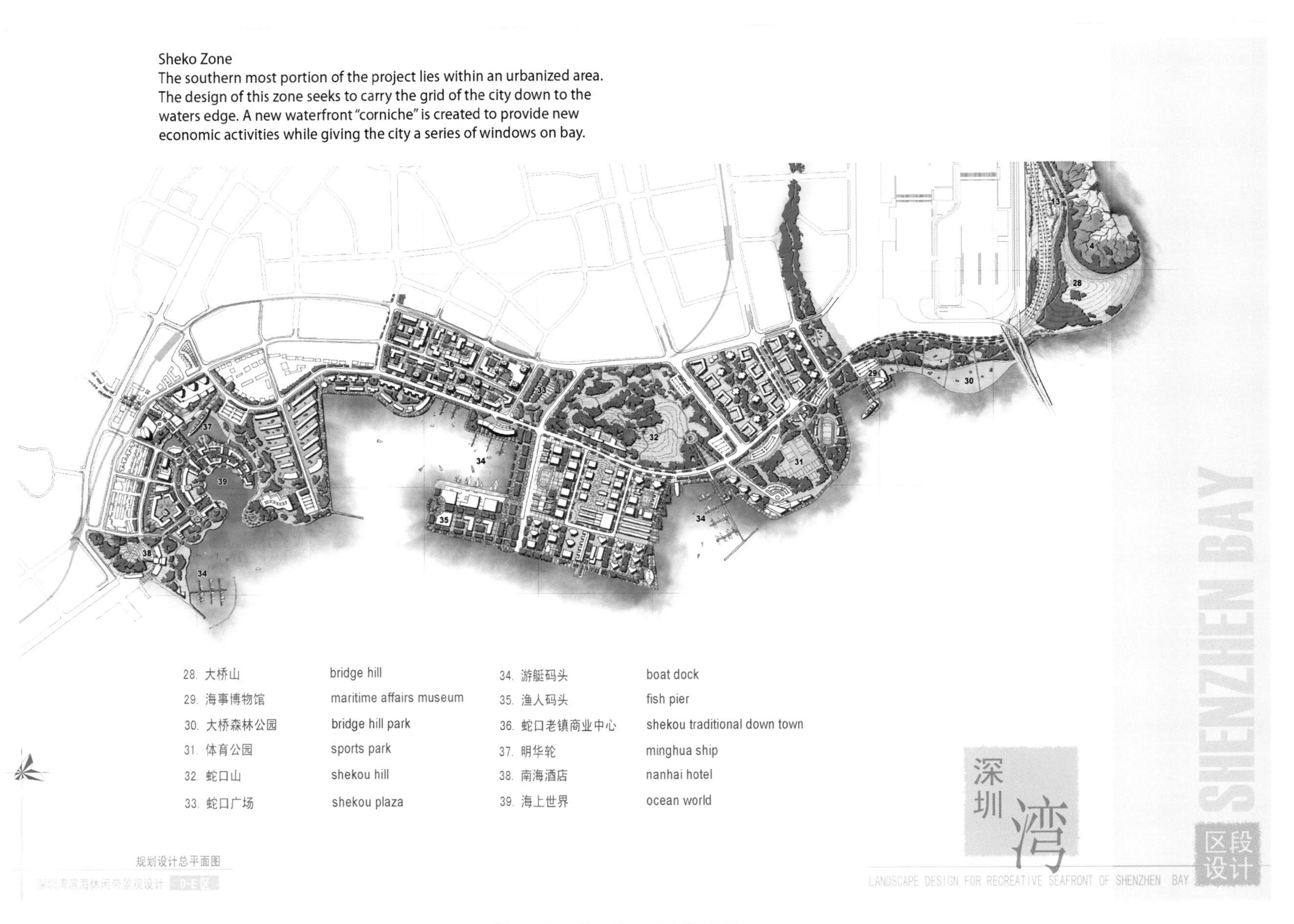


Shenzhen Bay Coastal Park Plan

On the western most portion of the project, a new community uses the existing street grid the inland city and connects it to the proposed reshaped shoreline. Pedestrian walks and bridges provide connections through the entire coastal zone, including an open-air bridge with unobstructed views connecting the north residential areas to the education center; a broad landscaped bridge connecting to the waterfront park and a bay edge walk with way-stations to provide shelter, rest and information along the recreation zones. On the west edge, a central plaza provides a view corridor from the land port and from the coastal boulevard for cars and includes a transit stop as well as steps that descend to the tidal zone and observation towers. At the south end of one mangrove restoration area, one of the proposed hills will include a forest, open meadows and a sheltered overlook. In an outer park there will be a large parking area and drive-by viewing stop, as well as a seaside promenade.

To provide open space for the urban residents, a large inland park with an inner bay has been planned. This bay will fuse the hardscape of the urban edges to the north with the more natural south edge and will include a tidal control structure to maintain the water level. The design of the park provides for islands, wetlands, woodlands along with formal lawns and informal meadows for a variety of activities. The plan encompasses an entertainment and education pavilion, formal gardens, an outdoor theater, restaurants and teahouses, along with convenient transit stops.

这个项目占地范围为 15 km，包括 3 个有特色的滨海区：西海岸、滨海湾和东海岸。总规划旨在将整个海岸线统一为一个紧密联系的整体。景观建设方案开始于大面积红树林的恢复和海草、盐沼植被的重建。

景观建筑师借鉴本国工程师的经验和专业知识来开发沿海区，并规划海岸线的保护和完善，如巨石边缘区。设计师建议在滨水区设置 4 座小山，以提供视觉参考点和海湾及城市的视野。

总体规划设计了一个海岸线教育中心，作为教育区和展览区，为学生、当地访客、游客和研究者提供滨水区物理、生物和社会环境方面的信息。这个教育中心除了室内设施，还包括户外沿海鸟舍、展示的潮汐池、教育长廊和教学馆，教育长廊会延伸到盐草区。

在项目区西部，大部分区域上一个新的社区利用现有的街道将内陆城区固定在网格中，并使之与提议改造的海岸线连接起来。人行道和桥梁联系了整个沿海区，包括一座连接北部居民区和教育中心的户外桥，提供一览无余的视野；一座连着滨水公园的景观桥；一个海湾边缘区的散步道，沿着娱乐区设有小站提供遮蔽，休憩和信息浏览。在西部边缘有一个中央广场，提供陆路口岸和海滨车道处的观景廊，还设有交通停车站、延伸至潮汐池的台阶和瞭望塔。红树林恢复区南部尽头设计的一座小山包括森林、开放草甸和一个隐蔽的瞭望台。 外部的公园中设计了一个大的停车场和过路观景点，以及一个海滨散步道。

为了给城市居民提供开放空间，设计师设计了一个包括内湾的内陆公园。这个内湾将北部城市边缘的硬景观与更倾向于自然环境的南边融合起来，也设有控制潮汐的建筑以维持正常水位。设计师在公园正式的草坪和非正式的草地旁设计了小岛、湿地和林地，为各种活动提供场所。交通停车站旁边设有一个兼具娱乐性和教育性的亭子、正式的园林、一个户外戏剧院、餐厅和茶馆。

# Doubletree by Hilton Avanos Hote

## 阿瓦诺斯希尔顿逸林酒店

Landscape Architect: DS Architecture – Landscape

Design Team: Deniz Aslan, Zeynep Emektar

Client: Dorak Tour Turizm

Location: Avanos, Nevşehir, Turkey

Area: 7,000 $m^2$

Photography: Gürkan Akay

Within a magical atmosphere, Hilton Avanos which is settled amidst 10,000 $m^2$ of lush green, in the sub province of Capaddocia, Avanos, is a significant tourism complex. The unique architectural style of the hotel block was revealed once more through the purity of landscape design that surrounds the settlement. The harmony between the architectonic structure and the landscape design was procreated by introducing a simple balance between hard and soft landscapes.

Rather than composing a detailed processed garden suitable to the cold and dry climate of Cappadocia, vineyards and poplar woods were proposed as the essential components of the landscape project. Almost located on a rural landscape, the project zone sheltered all the colors of the seasons; however, autumn was a little ahead of the rest.

The roadside facade of the vaulted structure of the entrance building was almost naked, if the linear niches were not taken into consideration. Consequently, in order to enrich this frontal silhouette, a strong calligraphic effect was created through a simple plantation texture.

阿瓦诺斯希尔顿酒店是一个著名的旅游中心，它置身于阿凡诺斯卡帕多西亚副级省的 10 000 $m^2$ 的绿地之中，气氛神秘。居住地周围景观的简单设计再次展现了酒店独特的建筑风格。软质和硬质景观的平衡设计使得建筑结构与景观设计之间互相协调。

这个景观项目的重要部分是葡萄园和杨树林，而不是组成详细处理的适于卡帕多西亚干冷气候的花园。 这个项目区域几乎是在农村，维持了四季的颜色；然而秋天要比其他季节明显一些。

如果没有那些线型的壁龛，入口处拱形建筑的路边门面几乎是没有装饰的。因此，为了丰富它的门面，一些简单的种植用来增加它的美感。

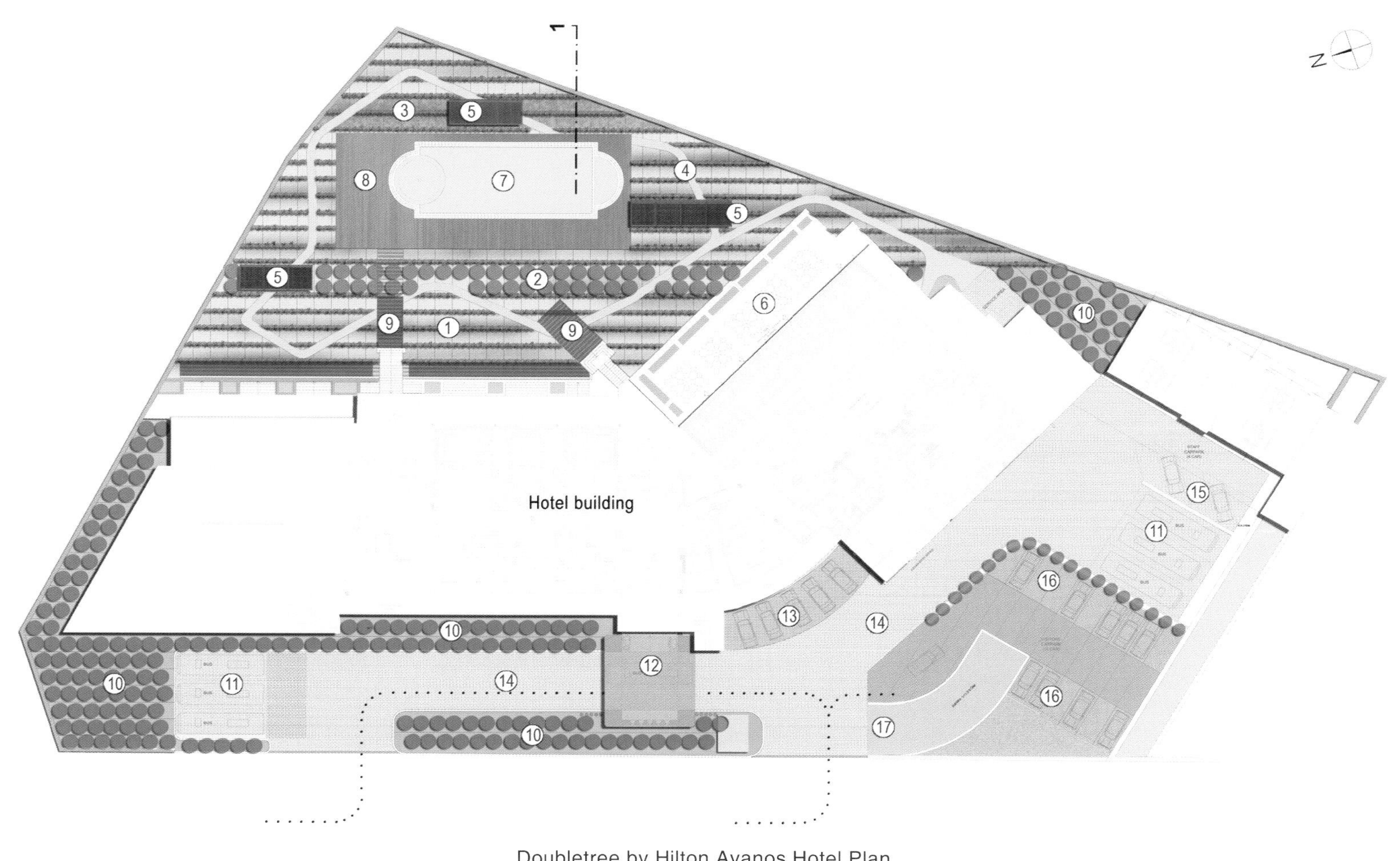


Doubletree by Hilton Avanos Hotel Plan

1. Vineyard
2. Fruit garden
3. Lavender Garden
4. Path
5. Bambusa Canopy
6. Service Area / Cafe
7. Swimming Pool
8. Sunbathin Wooden Deck
9. Wooden Platporm
10. Green Area
11. Bus Parking Area
12. Hotel Main Entrance
13. Car Park
14. Carway
15. Staff Carpark
16. Visitor Carpark
17. Underground Parking Entrance

# 31st Street Harbor

# 第31街港口

Landscape Architect: AECOM

Photography: Dixi Carrillo

The 31st Street Harbor transforms an underused portion of Lake Michigan shoreline in Chicago, Illinois, into a new marina and public park. Led by AECOM's landscape architects, the project melds a social and ecological improvement framework with a technical marine and civil engineering approach (also provided by AECOM), resulting in a vibrant gathering place for the neighborhood and regional visitors.

The landside development included a harbor services building/parking structure (which recently was certified LEED Gold) with a 5,853 $m^2$ accessible green roof, a great lawn, and enhanced waterfront public realm. A new fully accessible play area that connects the green roof area to the existing beach replaces a smaller, outdated playground.

The three large sail-like structures located on the green roof of the harbor services building serve as landmark for those on the water, cruising by on the newly configured bike and walking path or driving by the site on Lake Shore Drive.

The Chicago Park District's wider prairie program played well with the selected landscape plantings. This lakefront park is no longer just turf and trees, but rather incorporates low-water native coastal plantings that will reduce maintenance and irrigation demands, and provide food and cover for bird migrations through the city.

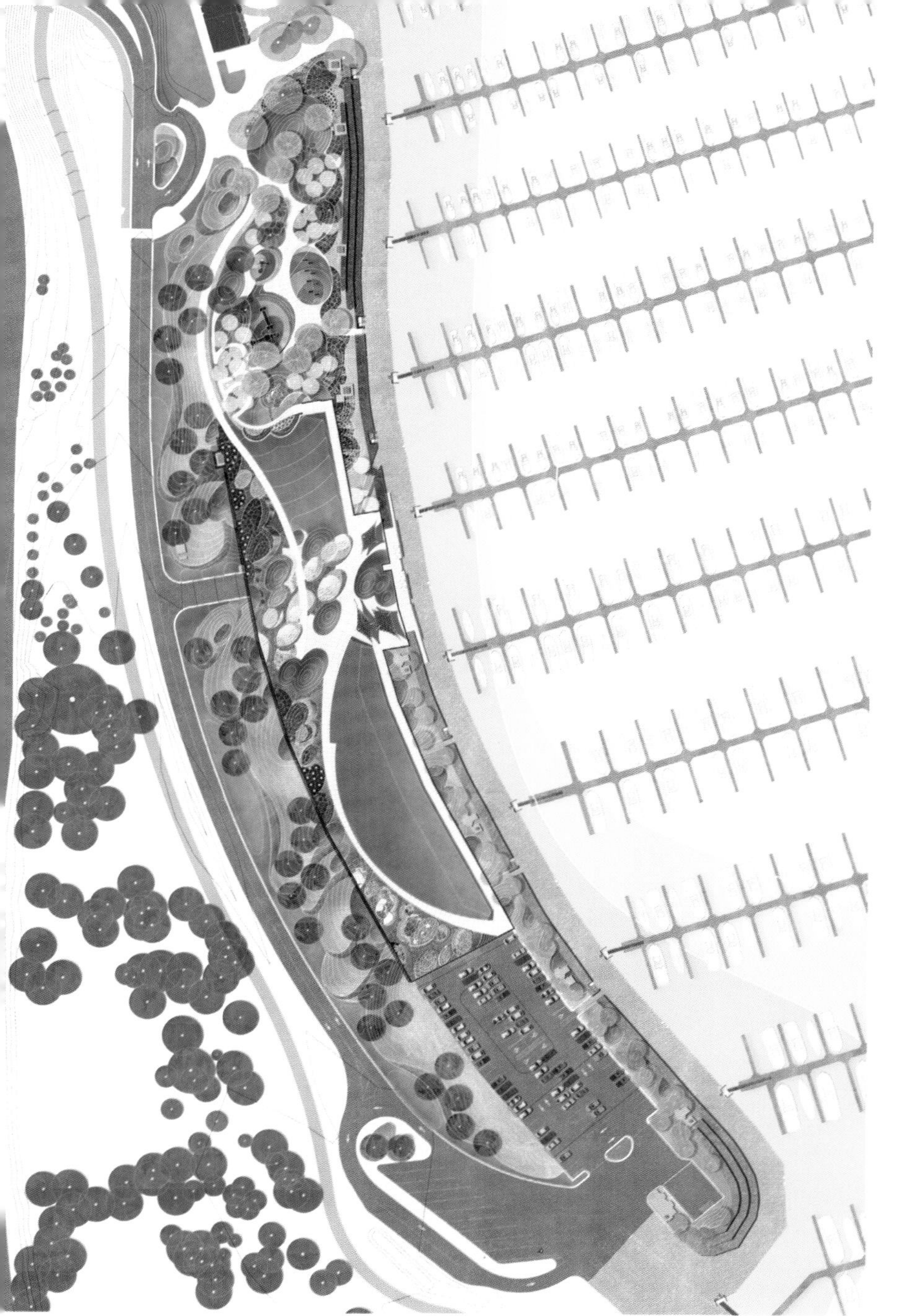

31st Street Harbor Site Plan

The climbing wall utilizes the vertical face of the garage to cater to varied climbing skill levels. It also provides an alternative means of access to the green roof.

The playground is a signature component of the development, providing a play and gathering place for a neighborhood excited for new parks and recreation activities. It features both fixed and flexible play space including rubberized dune mounds for children to interact with, a surprise water play area and a custom climbing wall which is built into the mounded grade of the park.

第 31 街港口将伊利诺斯州芝加哥密歇根湖岸线一处未被使用的区域转变成了一个新的码头兼公共公园。这个项目由 AECOM 公司的景观建筑师们领导，将一个社会及生态的完善提案融入到一个技术含量高的码头和土木工程方案中（也是由 AECOM 公司提供），为当地居民及整个区域的游客创造出一个充满活力的集会场所。

陆边地区的发展包括一个海港服务建筑 / 停车场（最近通过了 LEED 金级认证）、一个 5 853 m$^2$ 的可供使用的绿色区域、一个大草坪和改造过的滨水公共区域。一个可接近的将绿地与海滩连接起来的游乐区域取代了一个较小的已经过时的游乐场。

3 个大型帆状建筑物坐落在海港服务区的一块绿地上，那些在水上玩的、在新配置的车道兼人行道上穿梭的和在湖滨道上开车的，都视它为一个地标。

芝加哥公园区的大草原项目给选定的景观种植提供了很多便利。湖边公园不再是只有草皮和树木，它还包含低水位本土的海滨植物，能够减少维护和灌溉的需求，并且为城市里迁徙的鸟儿提供食物和庇护。

攀岩墙是车库垂直的那一面，满足各种攀爬水平的需求，它也是进入绿地的另一条路线。

游乐场是此项开发中颇具代表性的一部分，为居民们提供一个娱乐和聚会的场所。它的特色是游乐场所的布局既有条理又很灵活，包括供孩子们玩的橡胶沙堆、一个独特的水上游乐场和一个地势较高处定制的攀爬墙。

# Science Garden in Haifa Noble Energy Science Park at the Madatech Museum

# 海法科技园 马达科技博物馆诺布尔能源科技园

Architects: Mayslits Kassif Architects With Wachman Architects

Client: Madatech – The Israel National Museum of Science, Haifa

Location: The old Technion Court, Haifa, Israel

Area: 4,500 $m^2$

Photography: Mikaela Burstow

The Madatech – Israel's National Museum of Science, is located in the heart of a 28,328 $m^2$ campus in midtown historic Haifa. The two buildings on campus, designated for science exhibitions and education, were designed by Alexander Baerwald and date back to 1912.

Lying alongside the steepest street in Haifa, the park sprawls between these two landmark buildings, bridging their 5 m floor level gap with its undulating topography and serves as a continual link to the museum's indoors exhibitions.

Designed as an outdoor science museum the park juxtaposes a natural setup of trees and water with giant interactive exhibits using wind, sun and water, to demonstrate scientific principles.

The site's topography generated the principal of movement in the park, where visitors climb, almost unconsciously, through series of seven thematic courtyards, each focuses on the discovery of a noted scientist and descend back to the start point. The serpentines of waterfalls and dams, all dedicated to theme of water, link all courtyards together with an open-air amphitheater which serves for public assemblies and performances.

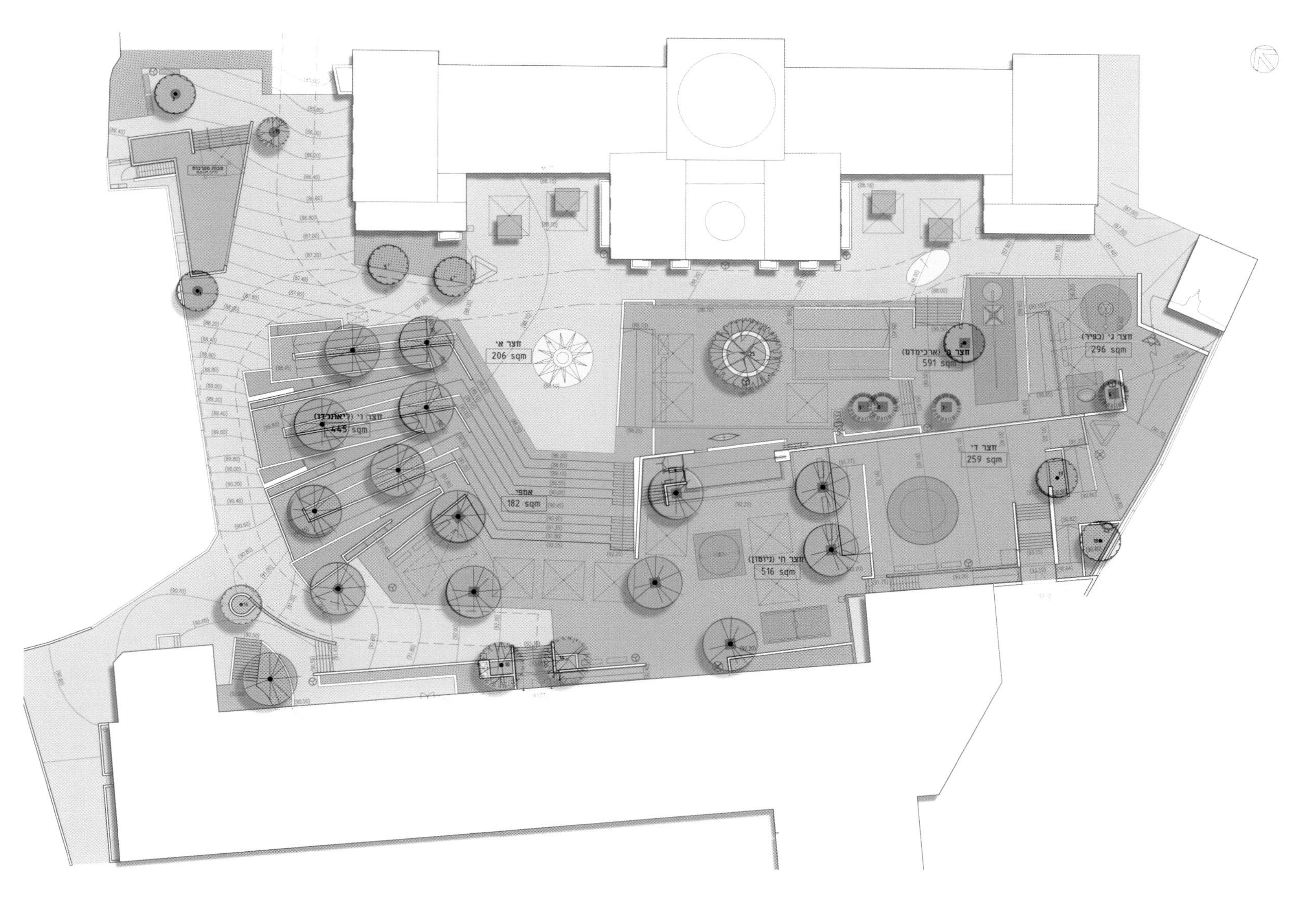


The Madatech Science Garden in Haifa Plan

马达科技馆是以色列的国家科学博物馆，坐落于一个占地 28 328 $m^2$ 的学校的中心，位于历史名城海法的市中心区。学校内两栋大楼用作科技展览和科学教育，由亚历山大 · 贝尔沃德设计，可追溯到 1912 年。

公园位于海法坡度最陡的街道旁边，从两栋重要的大楼之间伸展开，以其起伏的地形使两者在 5 m 高的地方连接起来。同时公园也为室内展览提供连续的引导。

公园作为室外科学馆，将树木、水这些自然景观与风能、太阳能、水能相互作用的大型展览串联起来，以彰显科学原理。

这里的地形决定了公园活动的主要内容，游客穿过一系列的 7 个主题庭院登高后，几乎都不自觉地专注于寻找著名科学家以及如何回到起点。蜿蜒的瀑布和大坝也强调了“水”这个主题，利用露天剧场将所有庭院聚集起来。露天剧场是公共集会和表演的场所。

# Trump Towers

## ▶ 特朗普大厦

Landscape Architect: DS Architecture – Landscape

Design Team: Deniz Aslan, Günseli Döllük, Elif Çelik

Client: Taş Yapı A.Ş+D Yapı A.Ş

Landscape Area: 13,300 m$^2$

Photography: Gürkan Akay, Cemal Emden, Günseli Döllük, Selda İpek

Trump Towers, one of the most significant investments located at the heart of business and urban life of Istanbul, is a unique project that has a smooth connection with the surroundings and is a natural part of the urban landscape through its car park and its connection to the subway.

The design which shapes both the urban and the private use-oriented outdoors and contains the conceptual design of the terrace and the courtyard of the towers, was based on a linear sequential setup that defined the utilization of all the gardens.

While designing the entrance square that forms an interface with the urban life, it was planned to create an urban area that attracts people to the building, and also that is used as a transition and a service area. The square became a meeting point with its catering platforms, its pool that matches the refraction of the ground, its green wall, which is also a symbol of eco-technology, that filters out the noise and dirt of outside.

An impressive garden was aimed for in the terraces at the high levels that is far away from the city's chaos. It was intended that the linear lines, on which the design is based, is to be felt most intensively in this garden, where several utilization possibilities were offered to the user as a multi-purpose open space. While one of the terraces was designed as a transition, walking and recreation area with the highest botanic intensity, the other terrace was designed as a garden carrying out the function of an open space and sunbathing area.

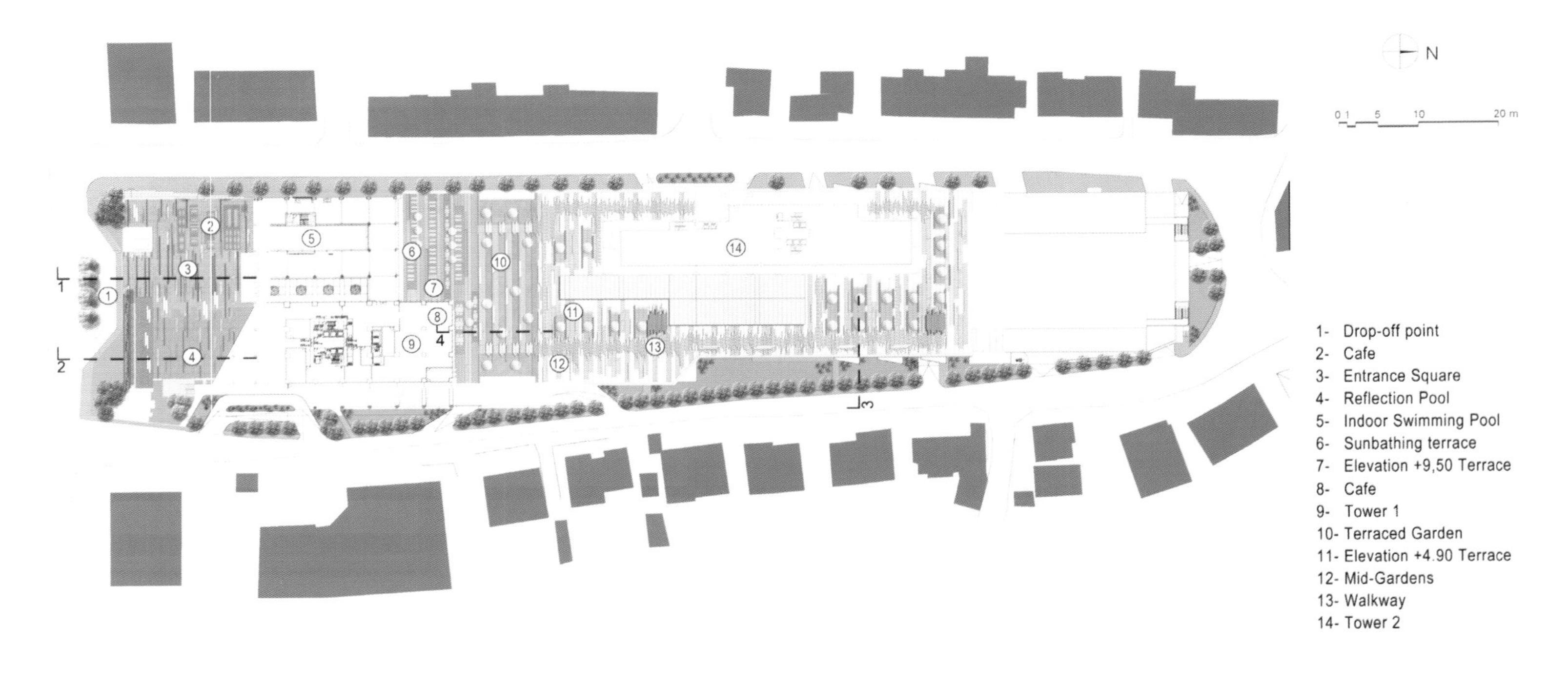
N
0 1 5 10 20 m
1- Drop-off point
2- Cafe
3- Entrance Square
4- Reflection Pool
5- Indoor Swimming Pool
6- Sunbathing terrace
7- Elevation +9,50 Terrace
8- Cafe
9- Tower 1
10- Terraced Garden
11- Elevation +4.90 Terrace
12- Mid-Gardens
13- Walkway
14- Tower 2

Trump Towers Detail Planter

TRUMP TOWERS MALL
SATURN
MIGROS
TOWERS
MALL

özgürce yaşa!
TRUMP
TOWERS

特朗普大厦坐落于伊斯坦布尔商业与城市生活的中心，是最重要的投资项目之一。它是一个独特的项目，与周围环境完美地融合，是城市景观中的自然部分，主要地点是停车场以及与地铁站相连的区域。

这个设计规划了城市公众和私人的户外空间，以明确园林使用的线性连续的组织为基础，包含对大厦阶地和庭院的概念设计。

设计师在设计与城市生活紧密相连的入口广场时，打算创造一个城市区域吸引大众到这里来，同时也可作为过渡区和服务区。广场上有餐饮平台和能反射地面的水池，也有象征着生态技术的绿墙，过滤掉城市的噪音与灰尘。这些设施使得广场成为了一个会议场所。

高处的露台上设计了一个令人倾佩的花园，使这里远离城市喧嚣。这个花园作为多功能的开放空间，提供多种用途，旨在使人们感受到一种强烈的线性元素，这也是此项设计概念的基础。植物最繁密的露台用作过渡区，供人们散步和娱乐。其他的露台被设计成花园，作为功能型开放空间，提供日光浴场所。

# Zhangzhou Green Lake Eco-park

## 漳州碧湖市民生态公园

Landscape Architect: L&A Design Group

Area: 2,400,000 $m^2$

Client: Zhangzhou Urban and Rural Planning Bureau,

Zhangzhou City Construction Investment and Development Co., LTD

Zhangzhou Bihu Citizen Ecological Park Concept Design

Zhangzhou locates in "Golden Triangle" district of Xiamen, Zhangzhou and Quanzhou with distinct location advantages. The Green Lake Ecological Park is defined as a top citizen water park among domestic municipal parks, reflecting ecological landscape planning concepts and public recreation characteristics.

The overall design principles:
Highlight the idea of ecological nature and low carbon environmental protection, respect local cultural features, inspire commercial vitality, carry forward community culture. Establish wetlands ecosystem to enable water as organic connection and bond of the entire park system and the realization of various functions, and also as the soul of the park. Regard the theory of ecological economy and tourism economy as our guidance, obey the principle of combining development and protection, integrate a variety of functions, provide people with places for tourism, vacation, leisure, education, culture and entertainment, making this eco-park become a green field of dreams for citizens.

Landscape design concepts:
Regard Chinese silk ribbon ,namely "dance show ribbon of ribbon" as the concept theme, showcase how to enhance the relationship between life and culture, urban and nature, to pass on the stories from past to future. Three ribbon areas around the park respectively represent natural landscape, urban landscape and cultural landscape, framing the content of this project. And of them relates to as well as highlights the bending water lines and the corresponding functional facilities and activities.

Function Division:
According to the planning layout, this eco-park could be divided into eight districts and two islands, each has its own feature while correlating to others. Eight districts: leisure area, lakeside business with sunshine lawn area, recreational activity area, exhibition area in the park, wetland landscape area, fishing leisure and science popularization education area, the riverside road landscape zone and cultural corridor of lakeside. Two islands: ecological island and narcissus island. The combination of each district and the adjacency relationship of different functional zones fully bent into the life of citizens and infused the natural atmosphere into the butted urban space.

Eco design:
The designers applied numerous ecological engineering and ecological infrastructure in the overall design, including constructed wetlands, wetland floating islands, rain garden, ecological open trenches, green roofs and walls,etc. The design aims to become the model of combining ecology and humanism through the use of various ecological infrastructures.

漳州处于厦漳泉“金三角”地带，区位优势明显。碧湖市民生态公园定义为国内市级公园中一流的、体现生态景观规划理念与公共休闲特色的市民水景公园。

总体设计原则：

突出生态自然与低碳环保的理念，尊重地方文化特色，激发商业活力，弘扬社区文化；建立湿地生态系统，使水体成为公园整个系统及各种功能实现的有机联系和纽带，成为公园的灵魂；以生态经济和旅游经济理论为指导，遵循开发与保护相结合的原则，将多种功能相结合，为人们提供旅游、度假、休闲、教育、文化娱乐的场所，使生态公园成为市民的绿色梦想之地。

景观设计理念：

以传统的中国丝绸缎带“生活之秀舞飘带”为概念主题，演绎如何加强生活与文化、城市与自然之间的关系，承延过去和未来的故事。三抹缎带环绕公园，分别代表自然景观、城市景观和文化景观，架构项目内容。所有设计都与选址的曲线形水管、相应的功能设施和活动相关并使其成为亮点。

功能分区：

根据规划布局，生态公园可划分为八区两岛相互关联各具特色的功能区域。八区：时尚休闲区、滨湖商业与阳光草坪区、康乐活动区、公园展示区、湿地景观区、垂钓休闲与科普教育区、江滨路景观带、湖滨文化长廊；两岛：生态岛、水仙岛。各区结合与城市不同功能区块的邻接关系，充分融入市民生活，并使自然气息渗透到对接的城市空间。

生态设计：

整体设计中设计师运用到诸多的生态工程和生态设施——人工湿地、湿地浮岛、雨水花园、生态明沟、绿色屋顶与墙面等。通过不同生态设施的使用，使其成为生态与人文结合的典范。

# St James Plaza

## 圣詹姆斯广场

Landscape Architect: ASPECT Studios

Architect: METIER3

Client: Juilliard Group

A rejuvenated urban plaza is the core to improving the retail and commercial viability of this renovated 1960's building.

ASPECT Studios were engaged to develop a design that assists with rejuvenating the public realm for the building, in line with the architectural refit. The landscape and building have a strong visual connection through the facade and paving treatments.

The planting design for the project includes mature Magnolias as the centre piece of the landscape. The Magnolias will provide assistance with wind mitigation whilst being a major visual node in the middle of both plazas.

The space includes formal seated areas for the retail spaces and informal areas for the local inhabitants to have lunch.

Location: Melbourne, Victoria, Australia

Area: 2,500 $m^2$

Photography: Andrew Lloyd

St James Plaza Plan

一个恢复活力的城市广场是提高这个 20 世纪 60 年代整修过的建筑的零售与商业广告生存能力的核心。

ASPECT 工作室参与了开发设计，以辅助根据建筑整修这栋建筑的公共领域。从建筑正面到地面铺砌之间，风景和建筑有很强的视觉联系。

这个项目的植物设计包括作为景观中心部分的成熟的木兰花。木兰花会帮助减轻风力，同时是广场中间的一个主要的视觉节点。

这里设有正式的为零售空间提供的就座区域，和非正式的供本地居民享用午餐的地方。

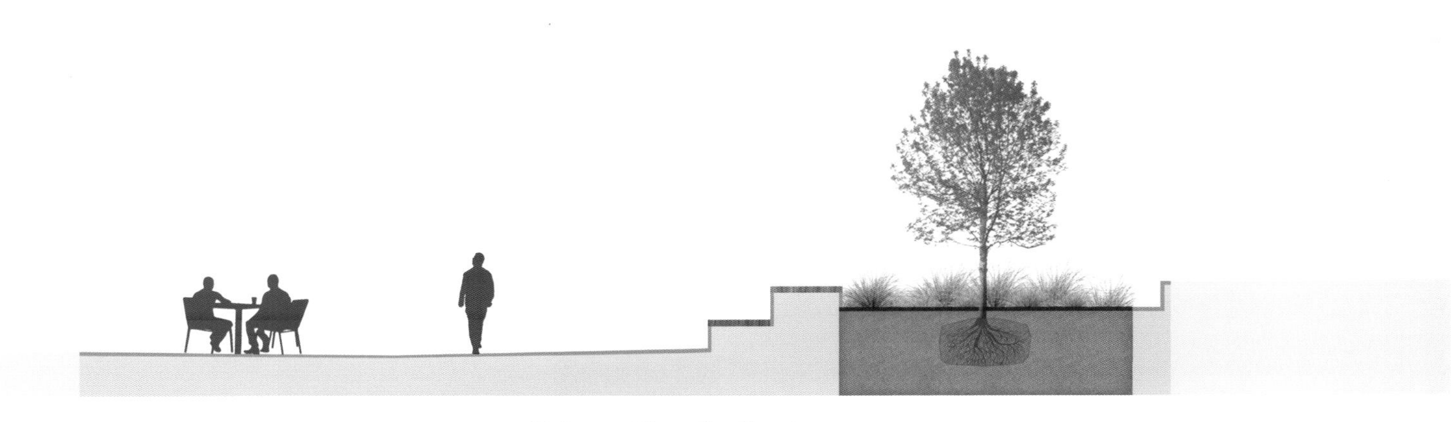

St James Plaza Section 1

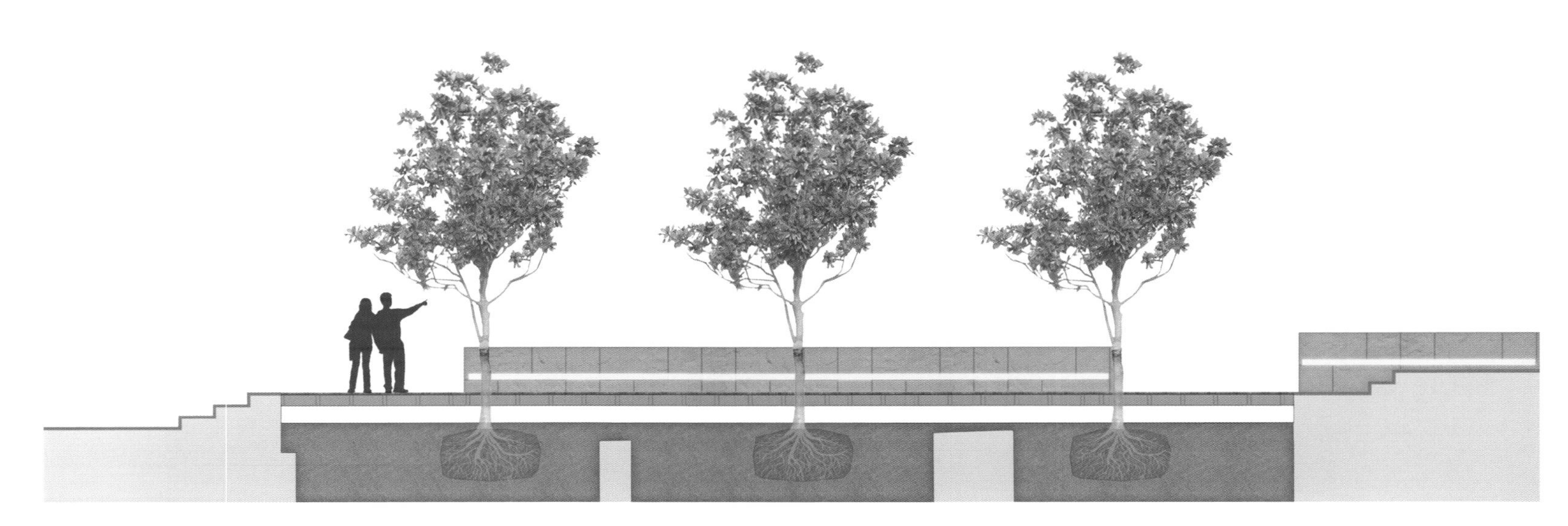

St James Plaza Section 2

# Cranbrook Junior School

## 克兰布鲁克小学

Landscape Architect: ASPECT Studios

Architects: Tzannes Associates

Client: Cranbrook Junior School

Location: Rose Bay, Sydney, Australia

Area: 13,400 $m^2$

Photography: Simon Wood

The Cranbrook Junior School was designed with the ambition of creating a "school in a park".

Working closely with Tzannes Associates, ASPECT Studios developed the landscape design to provide flexible spaces that best meet teaching, learning and play requirements.

The landscape consists of tilted turf planes, mass planted banks, timber and concrete seating pods and groves of tree planting. The large turf planes allow for the assembly of children and accommodates informal play during recess and lunch hours. The central courtyards have been designed to read as "green spaces" at the heart of the school.

Small "pods" surrounded by timber and concrete benches are etched out of the green spaces. The range of seating opportunities allow for small or large groups of children to congregate and play.

The school was required to provide vehicular access through the centre of the grounds — a challenge the design team resolved by creating a high quality paved shared road that is closed to vehicles during school hours when it becomes a plaza for the students to use.

The courtyards also double as potential learning areas that small classroom groups can gather.

The landscape design also included four tennis courts and re-orientation of sporting fields to the south of the main school.

克兰布鲁克小学的设计旨在创造一个“公园小学”。

澳派设计工作室跟 Tzannes 联合事务所密切合作设计了一个景观，即能最大程度地满足教学和游戏的灵活空间。

这个景观包括有坡度的草坪、植物密集的水岸、木质长凳、混凝土长凳以及一些小树林。课间休息或午餐时间，学生可以在面积较大的草坪上聚会，玩游戏。中央的场地名为“绿色空间”，位于学校中心。

座位荚被木质长凳和混凝土长凳包围，设置在绿色场地的外面。学校里有如此多可以就座的地方，这就方便了孩子们聚会和玩游戏。

学校要求有可以开车到中央场地的通道，设计团队设计了一条高质量的共享道路来解决这个困难。上课时这条铺砌的道路是学生的活动广场，靠近车辆。

这些场地同时可当做学生小组学习的地方。

这项景观的设计也包括主校区南面的四个网球场和重新设立的运动场。

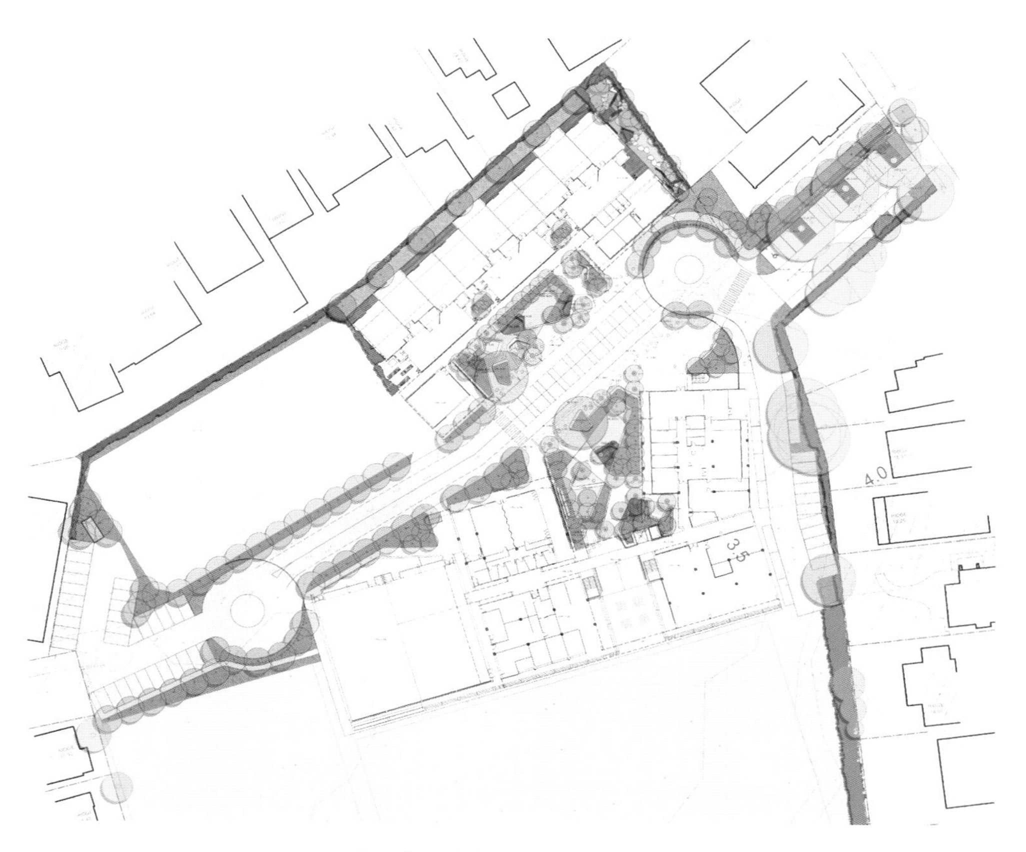


Cranbrook Junior School Plan

# Hornsbergs Strandpark

## ▶ 霍恩博格海滨公园

Hornsbergs Strandpark faces west to Ulvsundasjön and the evening sun. The waterfront and the three long floating piers give the visitor a feeling of floating into the light over the water. This is present particularly on hot summer afternoons when the park becomes an oasis for the surrounding residents and used for grilling and swimming. The park features several informal seating areas and a shower with a high seated tank for water heated by the sun that can be used by joggers.

The park is over 700 m long and consists of four parts. To the west lies a jetty for sunbathing with wooden docks jutting into the lake in different lengths. East of it is Kajparterren formed as a contrast to the organic Strandparken. It is a slightly raised horizontal disc slightly leaning towards the water. Far to the east is an existing part that has been renovated to be more accessible.

The project also includes the Moa Martinson square. For the proposed square design, Nyréns have focused on the spatial situation with a small spot at the edge of Ulvsundasjön and on the artistic adornment associations with the author Moa Martinson. Since the square surface is raised to provide access to the buildings it forms a difference of level with the street. The difference in level consists of a wall and staircase both possible to sit on. The stairs open to the square that is turned diagonally out to the lake with Kajparterren in the foreground. The warm evening sun falls in the same direction. The artistic decoration consists of large granite slabs that house engraved quotations from the books of Moa Martinson.

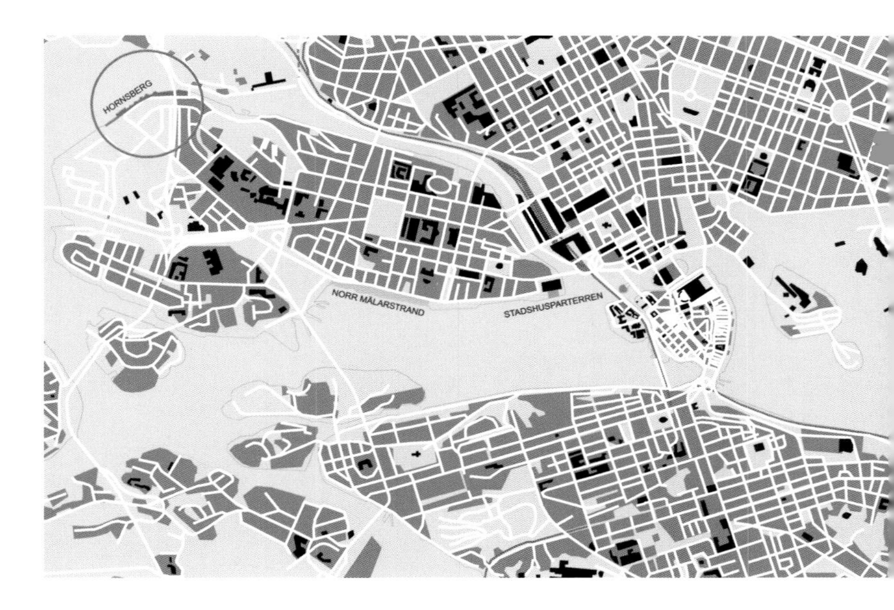


Landscape Architect: Nyréns Architects

Client: Stockholms stad Exploateringskontoret, Ewa Reuterbrand, Britt Mattsson, Fredrik Bergman, Torbjörn Engman, Monika Almqvist

Photography: Åke E: son Lindman

霍恩博格海滨公园向西面朝 Ulvsundasjön 和落日。海滨和 3 个长长的浮动码头带给游客一种漂浮在水上光辉之中的感觉。尤其在夏天的午后，公园成为周围居民的绿洲，供他们烧烤和游泳。公园中比较有特色的地方是非正式的座位区和一个可供慢跑者使用的太阳能热水淋浴区。

公园长 700 m，由 4 个部分组成。西部是可以享受日光浴的防波堤，长短不一的木质码头伸向水中。东部是 Kajparterren，与有机海滨公园形成对比。它是一个增加了水平高度的圆盘，稍微向水域倾斜。东部远处是已经存在的部分，经过重修后更容易到达。

该项目也包括莫瓦马丁松广场。对于这个广场的设计，Nyr é ns 将焦点放在空间设计方面即在 Ulvsundasjön 边缘的一个地点放置跟作者莫瓦马丁松相关的艺术装饰品。由于广场地面高度增加，进入大楼的通道的水平高度也相应开高，因此广场与街道之间形成了水平差。这种水平差由围墙和阶梯构成，可供人们坐下小憩。阶梯向广场开放，在水域的斜对角方向，前面是 Kajparterren，这里也是日落的方向。艺术装修包括室内雕刻有 Moa Martinson 书中的大面积花岗石板。

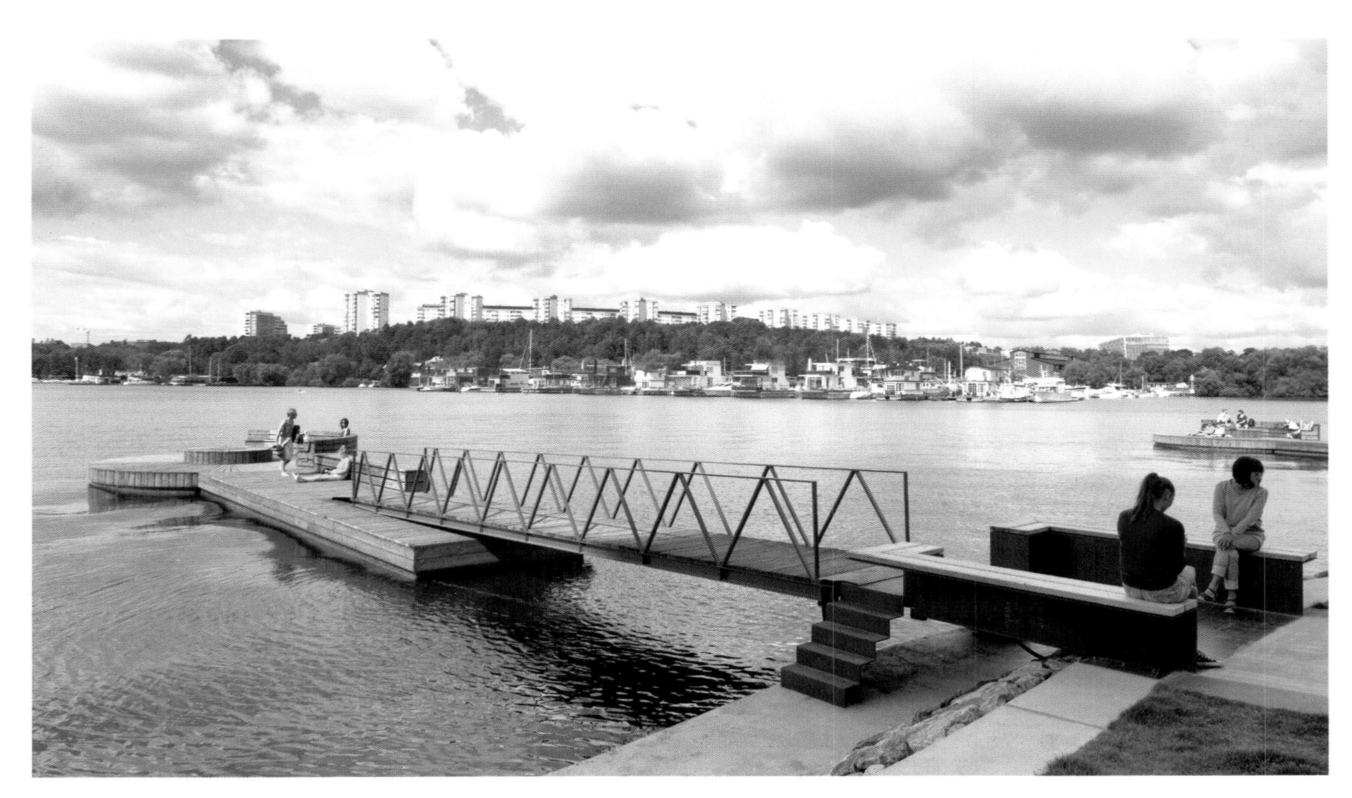

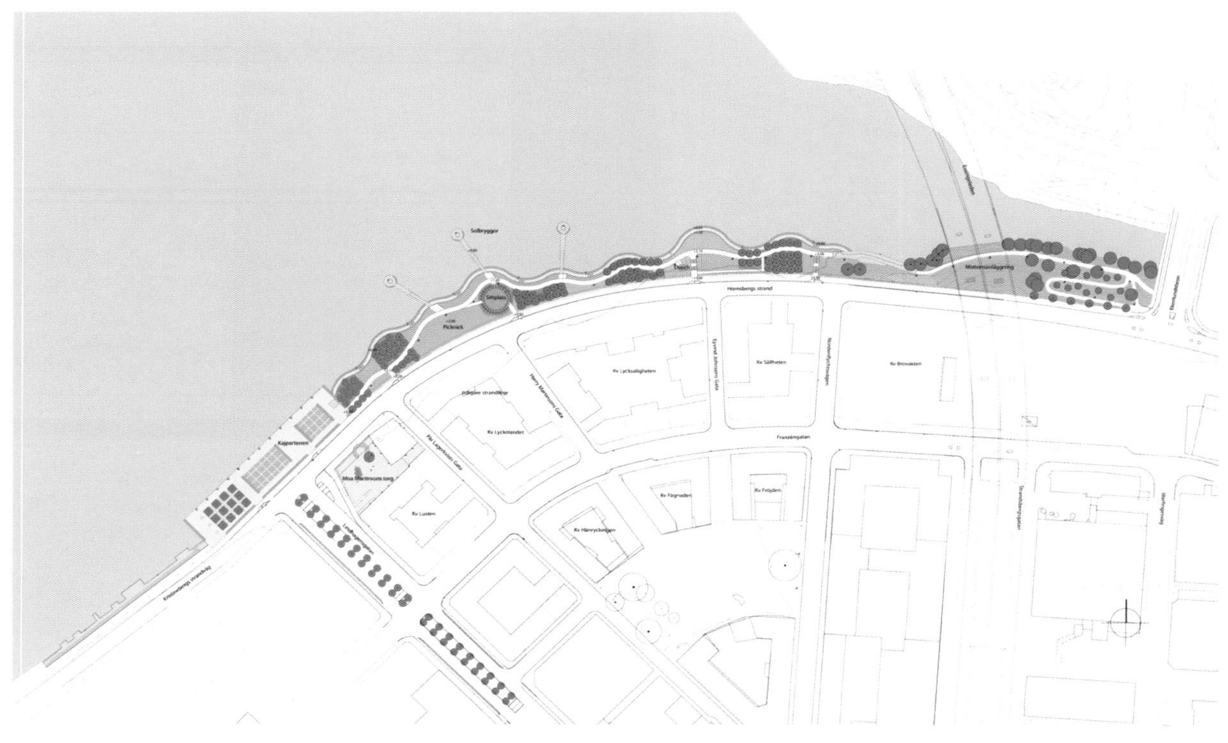


Hornsbergs Strandpark Plan

han super inte och springer inte med andra
fruntimmer
det är ändå något, sa mor
barnen hade fått tillåtelse
att hålla skymning i köket

# Blokhoeve, Nieuwegein

## ▶ 尼沃海恩布罗克霍夫社区公园

Landscape Architect: Dijk & Co

Playground and Skatepark Design: Carve

Location: Blokhoeve, Nieuwegein (NL)

The Park "island-West" is the community park of the new district Blokhoeve. The park was designed by Dijk & Co landscape architecture. It is a hilly grassland with loosely scattered trees. Although it is a new park, the trees are quite old, several years ago the existing hungarian oak and linden trees were replanted, stored and placed back in the park after the construction period. The pathsystem is designed as a "liquid gel" that connects all entrances and includes the sports and playgrounds as a logical but playful pattern. The old location contained an old running track that was integrated in the new design. It now is a fresh running and skating track, that encloses the new sports and play functions.

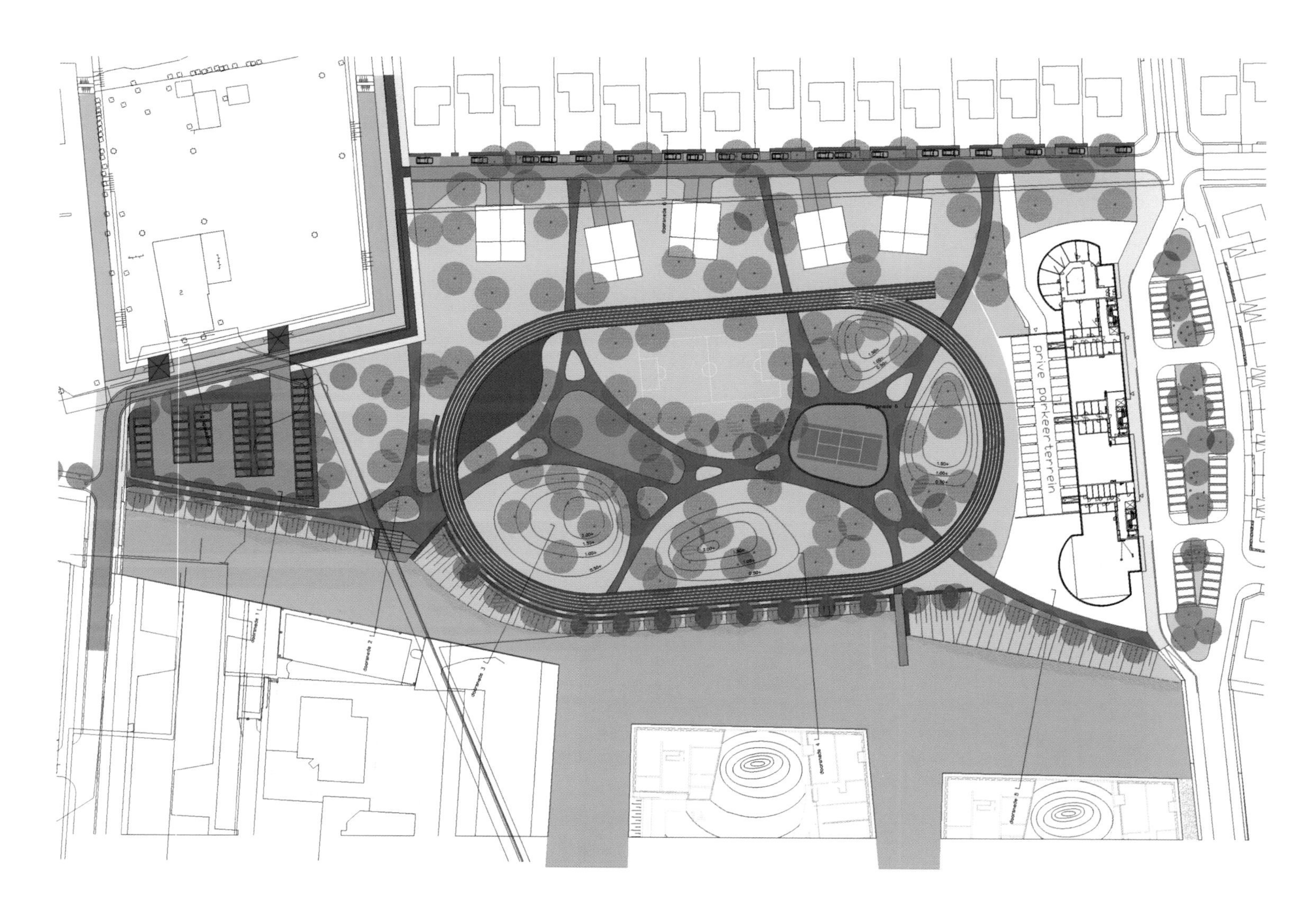


Plan Park eiland-West

The skate and play landscape was designed by Carve. The skate objects "stick" to the inside edge of the track, and are made of light coloured concrete that forms a nice contrast with the dark asphalt of the track. Being a neighborhood park with a sporty character, the choice was made to integrate game and sport in the play objects, too. The play object, with its vertical tree trunk forest, balances the vertical direction of the mature trees in the park. The play cubes, which hang tilted between trunk forest, are designed from the scale and perception of the child. The exterior of the same play objects, however, is a professional boulder wall, so parents are challenged to climb on it. The playcubes are placed quite high to prevent small children to climb on the outside. Nevertheless, children can crawl and climb from one cube to the other through elevated climbing paths. Two worlds are united in one object: they don't blend but they do meet. With this, the children's playground also becomes a playground for adults.

由 Dijk & Co 景观建筑公司合作设计的西岛公园是 Blockhoeve 新区的一个社区公园。公园内的树木零星分布在小山的草地上。虽然这是一个新的公园，但是里面的树木都是古树。几年前，原来在这儿的匈牙利橡树和菩提树被移走，施工期过后，它们又回到了公园的怀抱。道路系统像液态凝胶一样连接了所有的出口，并且以富有逻辑而不失幽默的形式环绕着体育场。体育场旧址包括一个古老的跑道，这个跑道被整合在了新的设计里，成为现在的新跑道兼滑冰场，它环绕着新的运动场和娱乐区。

滑冰场和娱乐区是由 Carve 设计的。滑冰场紧挨着跑道的边沿，它是由浅色的混凝土制成，和跑道的沥青构成了鲜明的对比。作为一个运动型的社区公园，设计的选择也要整合比赛与运动。运动场的树林应与公园里的古树协调一致。从孩子的比例和感觉方面考虑，立方体形状的体育设施被倾斜地悬挂在树干间。此外，运动场的外围是专业的大卵石墙，因此，家长们很难爬上去。体育设施也被放置得非常高以防小孩子爬出去。然而，孩子们可以通过架高的攀爬路径从一个体育设施爬到另一个。两个区域合成一个区域，两者不混合并且形成互补。这样，公园既是孩子的游乐场也是大人们的运动场。

# The Hive Worcester Library Landscape

## ▶ 蜂巢状伍斯特图书馆景观

Landscape Architect: Grant Associates

Architect: Feilden Clegg Bradley Studios

Client: Worcester City Council

The Hive is Europe's first joint university and public library — a unique academic, educational and learning centre for the city of Worcester and its university. The "BREEAM Outstanding" project was designed by architects Feilden Clegg Bradley Studios with a distinctive and sustainable landscape design by Grant Associates.

Grant Associates' landscape design brief was to create a high quality landscape environment that would become a distinctive and exciting visitor attraction — a place which would capture a sense of history and place whilst reflecting on the contemporary themes of sustainability and technological innovation.

The landscape is based on a strong narrative derived from the local landscape of the River Severn, Malvern Hills and the Elgar trail that inspired Land of Hope and Glory and key storytelling themes:

· Nature uplifts the spirits — the landscape spaces are arranged to "enlighten and delight" inviting visitors to experience the therapeutic qualities of nature, an encounter with birdsong, scented plants, colourful wildflowers and dragonflies.

· Healthy water for sustained life — demonstrates to visitors the importance of healthy water for life and the ability of natural systems, not man made chemicals, to take care of this.

· Knowledge and Heritage — creates a special sense of place derived from the primary circulation route the causeway.

The 2 hectare site comprises a series of islands and belvederes overlooking two landform basins containing rich local damp meadow and the causeway, an extension of the city wall route, routes circling around and through the centre. Highlights include: the water meadow, sustainable drainage system, habitat islets and the causeway.

蜂巢状伍斯特是欧洲首家联合公共图书馆，也是伍斯特大学学术、教育和学习的中心。“环境评估优秀”项目由 Feilden Clegg Bradley 工作室的建筑师们设计，包括一个由 Grant 事务所设计的与众不同的可持续发展的景观。

Grant 事务所的景观设计梗概是创造一个高品质的景观环境，使之成为一个与众不同的令人振奋的景点，承载着历史意义又同时体现出可持续发展和技术革新这些当代主题。

这个景观基于对当地塞文河、摩尔纹山脉和埃尔加路这些景观的强烈叙述性推断，因为它们激发了希望与荣耀的灵感，展现了关键的故事主题：

· 自然振奋精神——景观空间的安排是为了启发和愉悦游客。这里有鸟鸣，有芬芳的植物，有五彩斑斓的野花，还有蜻蜓，带给人们一种心旷神怡的享受。

· 健康和持续的水——向游客展示健康水系的重要性以及自然生态体系而非人工化学品的能力，提醒游客珍爱大自然。

· 知识和遗产——在长堤主要环线上营造一种特殊氛围。

这个2公顷的地方由3个部分组成：一个群岛系；一系列观景楼，俯瞰着两个长满草甸的盆地；长堤，城墙路线的一个延伸，盘旋至市中心。重点是湿草甸、可持续的排水系统、栖息岛和长堤。

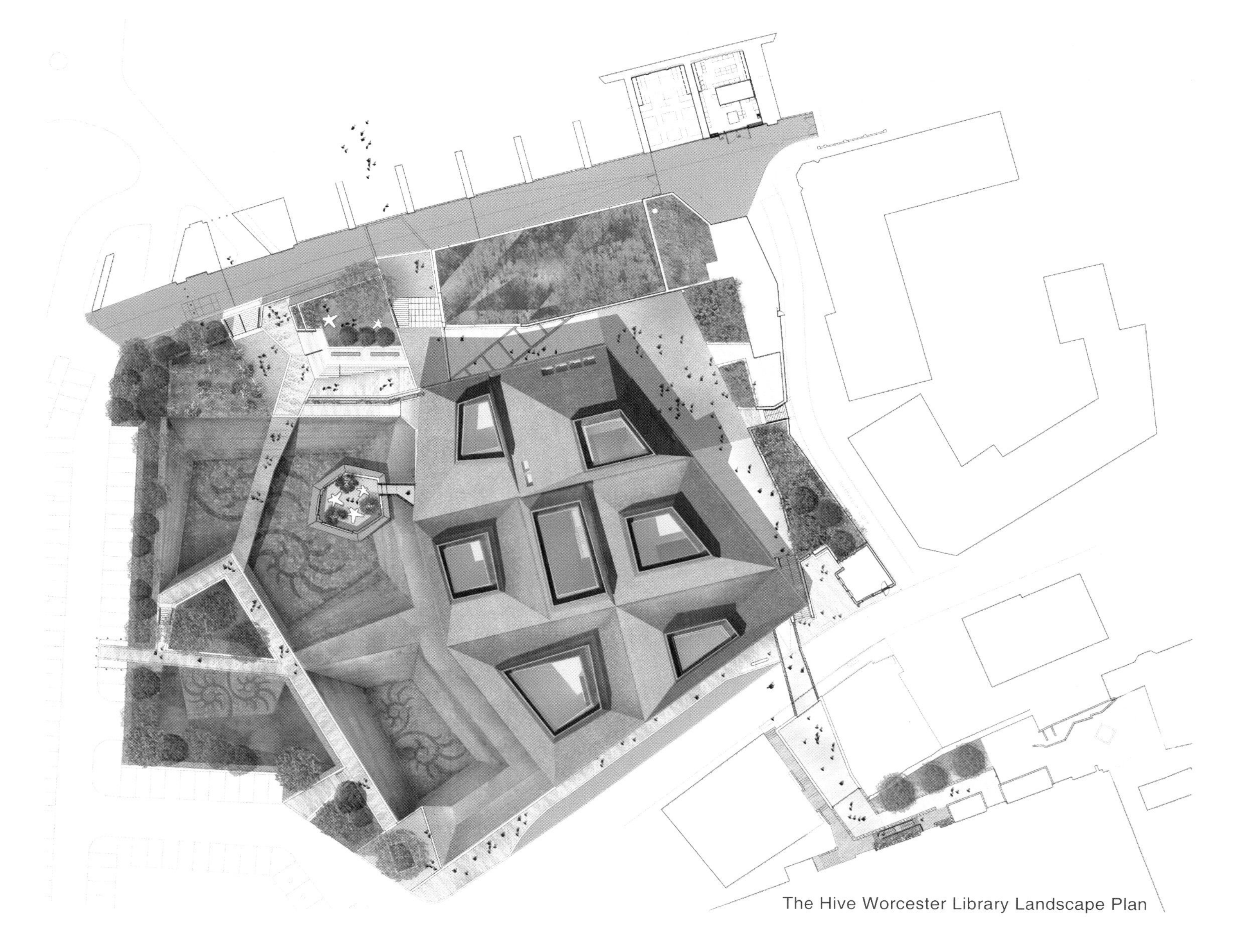

The Hive Worcester Library Landscape Plan

# Rochetaillée

## 罗谢泰莱埃

Landscape Architect: In Situ

Area: 60,000 $m^2$

The Rochetaillée site is exceptional for several reasons: it borders the river Saône on its left bank for over 2 km and follows a generous convex curb. Because of this, these shores are not subjected to violent currents and the location creates a landscape which opens up to the horizon and offers multiple points of view.

The project of the Saône riverbanks encompasses the river from the town Rochetaillée to its greater neighbour, Lyon. In the sequence of the project for which our team was selected, the shores are the widest and the orientation is west-south-west rather than north-south as is the case further down the river.

The design is extremely simple. In Situ have mainly worked on earthworks in order to strengthen the flood-damaged banks and to reveal a broad prairie, a green beach for the Saône. This generous space can host events, concerts and various performances. Here they have recovered air, wind, the horizon, trees and modest facilities: picnic tables and wide lounge benches for the afternoon nap. The popular "guinguettes" which are open-air cafés, punctuate the long walk along the river where pedestrians and bicycles have found their rightful place.

Four works of art have been designed specifically for the site by artists with whom have collaborated: Le Gentil Garcon, Didier Fiuza Faustino, Lang & Baumannet Tadashi Kawamata.

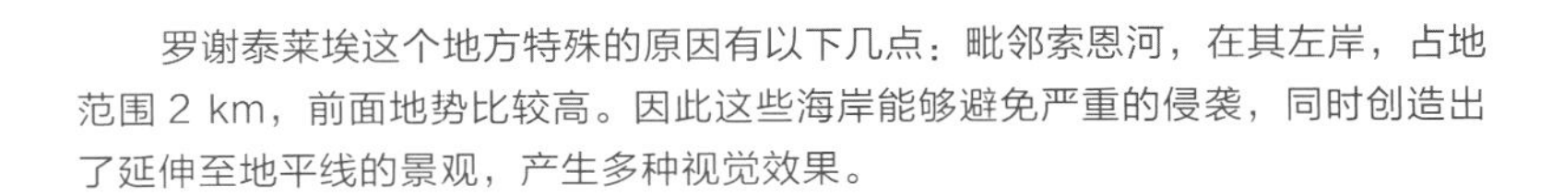

罗谢泰莱埃这个地方特殊的原因有以下几点：毗邻索恩河，在其左岸，占地范围 2 km，前面地势比较高。因此这些海岸能够避免严重的侵袭，同时创造出了延伸至地平线的景观，产生多种视觉效果。

索恩河河岸的设计范围是从罗谢泰莱埃镇到里昂市。Insitu 设计团队针对的一系列项目中，河岸的范围最广，方向是西南西，而不是地势较低的南北方向。

设计方案十分简单，在原地设计了很多工程来巩固被洪水损坏了的河岸，同时恢复一片植被，作为索恩河的绿色沙滩。这个广阔的地方可以举行大型活动、会议、不同类型的表演。他们恢复了这里的很多条件，比如空气、风、地平线、树木和一些便利设施，即野餐桌和宽敞的长椅，供人们午休使用。那些受欢迎的露天咖啡馆供河边行人和骑行者途中休息。

艺术家 Le Gentil Garcon、Didier Fiuza Faustino、Lang & Baumannet Tadashi Kawamata 跟设计团队合作针对这个地区详细地设计了 4 个艺术品。

# Eye-plaza

## EYE 广场

Landscape Architect: LANDLAB studio voor landschapsarchitectuur (Arnhem)

Client: City of Amsterdam and Projectbureau Noordwaards

Location: Amsterdam, the Netherlands

Photography: Mark Hell

Just north of Amsterdam's central station the City of Amsterdam is developing a new high density urban district called Overhoeks. Separated from the inner-centre of Amsterdam by the river IJ this new district needed an architectural icon that links Overhoeks to the IJ and Amsterdam's historical city centre. In this icon, the new film museum — the EYE — was to be accommodated. The Austrian bureau Delugan Meissl Associated Architects won the competition with a unearthly, eye-catching design. Due to safety regulations the building could not stand directly to the water's edge. A design for the transitional zone between the museum and the IJ had to be made.

In 2008 LANDLAB was asked to design the surroundings of the museum. LANDLAB made a purposely reserved design to avoid any competition with the building's shapes and to make the transitions between the museum, the plaza and the IJ as smooth as possible. At the water's side of the museum LANDLAB designed a large, sober plaza consisting out of a series of lazy sloping concrete plateaus. The interplay of these plateaus results in a variety of edges which invite people to sit, play and enjoy the view. The rear of the museum is set in green to establish its relationship with the Oeverpark and the still to develop district.

Since its opening in 2012 by the Dutch queen the EYE attracts many visitors that thankfully use the plaza to promenade, enjoy the sun and scenery.

阿姆斯特丹中央车站的正北部在开发一个新的名为 Overhoeks 的高密度城区。IJ 河将这个新城区与阿姆斯特丹的内部中心隔开。新城区需要一个标志性的建筑来使 Overhoeks 与 IJ 河及阿姆斯特丹的历史名城的市中心连接起来。因此新的 EYE-film 博物馆成为了目标。奥地利 Delugan Meissl 建筑事务所以其神秘而耀眼的设计赢得这次设计权。出于对安全性的考虑，这个建筑不能直接屹立于水边。博物馆与 IJ 河之间的传统地区需要进行设计。

2008 年，LANDLAB 应邀设计博物馆的周围环境。LANDLAB 做了别具匠心的设计，来避免建筑外形上的不妥，以尽量完成博物馆、广场和 IJ 河之间的完美过渡。博物馆旁的水岸附近设计了一个大型而庄严的广场，由一系列的混凝土斜坡组成。这些斜坡的相互作用产生的各种边缘区域，可供人们休息、玩耍和欣赏风景。博物馆后面是绿地，目的是建立与 Oever 公园以及在建地区的联系。

自从 2012 年荷兰女王开放了 EYE 广场后，它吸引了很多游客满怀感激地来这里散步，沐浴阳光和欣赏美景。

EYE-plaza 3D Rendering

over ◆ hoeks
Make

# Park Killesberg

## 基乐斯山公园

Landscape Architect: Rainer Schmidt Landscape Architects + City Planners

Associate Architect: Pfrommer+Roeder Landscape Architects

Client: State Capital City Stuttgart

Site Area: 100,000 m²

Photography: Raffaella Sirtoli, ARGE Zukunft Killesberg

The design is conceived as the interweaving of two themes that mark the Killesberg district: a soft landscape close to nature, and man-made quarries as hard topographies. The result is a landscape that tells its own story. The hard karst forms of a quarry topography, as though chiseled out, change typically over many years. Out of the sharply broken material the forms are rounded off until they become a soft landscape covered with earth and green. This theme within the meadow topography unites the three areas, Feuerbach Heath, Green Joint and the Park before the Red Wall. The result is a soft park topography of meter-high meadow cushions laid upon these areas, joining them together in harmony.

All areas together express the concept of sustainable and ecological development. The resulting roof water is collected in a huge underground cistern in the existing former exhibition building, and piped back into the park's lake and the natural water cycle. The lawn pillows of the park form various habitats for flora and fauna by their different microclimatic conditions. The landscape of the park extension interlocks with the adjacent residential area "Forum K", whose detached houses open to the park so that the green spacescan infiltrate into the center of the new district.

这个设计融合了基乐斯山的两个标志性主题：亲近自然的软景观和地形恶劣的人造矿场。因此这个景观其实就是在介绍当地的实际情况。喀斯特地貌形成了一个矿场，仿佛是被凿出来的，其外形随着时间而慢慢改变。在大幅破碎的喀斯特地形外，有一处被精心修圆磨光过的景观：绿地软景观。草原地形将以下三个区域连成一体，即 Feuerbach Heath、Green Joint 和 Red Wall 前面的公园，形成了一个绿色的公园地貌，像是这些区域之上 1 m 高的草垫，使它们协调地联合起来。

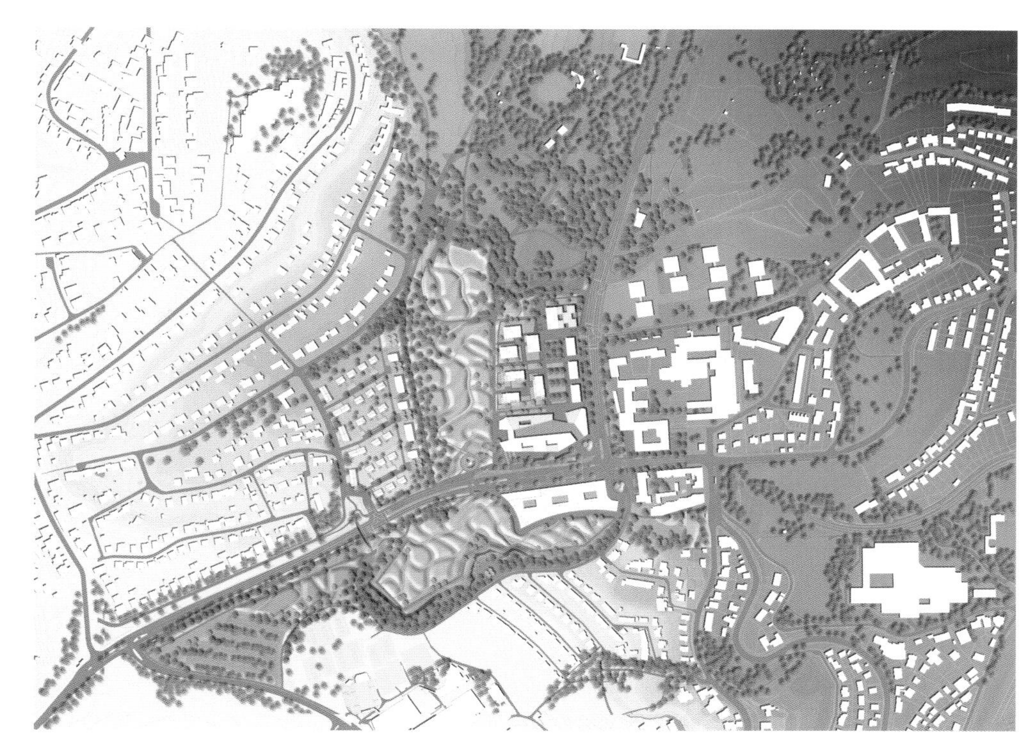

Park Killesberg Plan

所有区域都坚持了可持续的、生态的发展观念。屋顶的集水工作通过以前的展览大楼地下的大水箱进行，然后被输送到公园的湖中及自然水循环系统中。公园里的草坪以其不同的微气候，为植物和动物提供了多样化的栖息地。公园景观与相邻的居民区“K 论坛”互相影响，密切相关。居民区里分散的房子面向公园和绿地，因此居民必须穿过这个新区的中心。

This project celebrates the active participation of 11 abandoned rail line tracks of well-loved trains that slowly lumber through this downtown on a viaduct. This 76,890 m$^2$ urban park occupies the seam historically created by a major 4.6 m high rail viaduct that bisects the downtown. The park is formed by a new topography that carves the site for a lake and stream, providing flood protection and biofiltration. A range of knolls allows viewers to experience the train traffic first hand, creating a "trainfront" park. Land forms also give shape to a range of festival and performance spaces in this food and music-loving city.

A former warehouse and brick-making site, much of the park is formed with materials recovered from historic uses to create multi-functional plazas and paths that incorporate performances, food trucks and spaces to relax. The park is 4 blocks long by 1 block wide and was historically, the lowest point in town. The scheme draws on this ample water in creating a large reservoir for irrigation which also discharges through a stream and series of ponds as a summer fountain. Needed floodwater storage is created by excavating for this water system, using the spoils to create a series of knolls along the rail viaduct. The park contains performance venues of varying scales from small to extra large such as the annual "Crawfish Boil" attracting 30,000 music fans. Noisy or quiet, day or night, the park is only completed by the industrial ballet of freight cars slowly rolling in both directions.

# Birmingham Railroad Park

## 伯明翰铁路公园

Landscape Architect: Tom Leader Studio

Client: City of Birmingham, Railroad Park Foundation

Location: Birmingham, Alabama, USA

Area: 76,890 $m^2$

Photography: Jeff Newman, Sylvia Martin, Bradley Nash Burgess

这个项目得益于十一条被荒废了的铁轨，之前这条轨道上的火车备受人们的青睐，在高架桥上缓慢驶向市中心。这个占地76 890 $m^2$的公园，包括一个裂缝区，由将市中心一分为二的4.6 m高的主要铁轨高架桥造成。公园由一个新的地形构成，包括湖和小溪，具有抗洪和生物过滤的作用。一系列的小山创造出一个火车边的公园，使观众能够直接体验铁路交通。该地形也为这个热爱美食和音乐的城市规划出一系列节庆和表演的空间。

公园曾经是一个仓库和制砖场，它大部分利用之前建筑的回收材料，来创造多功能的广场和道路，为演出、食物运输和休息提供场所。公园有4个街区长，1个街区宽，所在处为城市海拔最低点。这个项目利用富裕的水资源建成一个大面积的灌溉蓄水池，并将水源源不断地排入公园内，形成溪流和池塘。水域系统中开凿了不可或缺的蓄洪池，多余的土料堆积成铁轨高架桥旁的一系列小山。公园包括不同规模的表演场所，超大规模的有年度“Crawfish Boil”音乐节，吸引了30 000名音乐迷。不论喧闹或宁静，白天或晚上，两侧缓缓行驶的货柜火车用一支“工业芭蕾”完美地诠释了场地的精神。

AERIAL PLAN

1 CRAWFISH BOIL STAGE
2 LAWN TERRACES
3 WEST GATE PLAZA
4 POND
5 STREAM
6 SKATE BOWLS
7 POND
8 TODDLER PLAY
9 RAIL TRAIL
10 STROLLING GARDENS
11 GREEK THEATER
12 BIRCH BOWL
13 LAKE
14 RAIL TRAIL BRIDGE
15 WETLAND
16 AMPHITHEATER
17 EAST GATE PAVILION
18 INTERMODAL STATION
19 AMTRAK STATION
20 CULTURAL FURNACE PROJECT
21 DOWNTOWN
22 DOWNTOWN BALL PARK (UNDER CONSTRUCTION)

Coke

NS

# Umeå Campus Park

## 于默奥大学校园

Landscape Architect: Thorbjörn Andersson with Sweco architects

Design Team: Staffan Sundström, Emma Pettersson, Mikael Johansson

Client: Umeå University

Location: Umeå, Sweden

Area: 23,000 $m^2$

Umeå University is a young university, founded in the late 1960s. Here, ca 35,000 students from all over the world study in all fields of knowledge. Umeå University is located by the coast, approximately 300km south of the Polar Circle.

A campus park should supply with a variety of designated places with the capacity to host informal discussions and exchange of ideas. It is in the open, non-hierarchical spaces, rather than in lecture auditoriums or at laboratory microscopes that the truly creative interaction between students, researchers and teachers occurs. The quality of the campus park thus enhances the attractiveness of the university as a whole.

The new Campus Park at Umeå University consists of 23,000 $m^2$ sun decks, jetties, open lawns, walking trails and terraces organized around an artificial lake. An island in the lake is the point of departure for a small archipelago with bridges leading to the southern shore. Here, the visitor meets a hilly landscape with sunny as well as shaded vales, interspersed by the white trunks of birch trees.

In front of the lively Student's Union, an outdoor lounge is laid out in the direction facing the sun. The lounge is a series of gravelled terraces in fan-shape, each terrace having café furnishing and shaded by multi-stemmed trees. The new Campus Park is the result of a competition, held in 2007.

于默奥大学是一所年轻的大学，成立于 20 世纪 60 年代末。这里大约有 35 000 名学生，他们来自世界各个国家，学习各个领域的知识。于默奥大学坐落于海边，离极圈约有 300 km。

大学校园应该提供各种指定的场所，能够举办非正式的讨论会和交流会。它处于一个开放的非层次的空间，而不是在演讲课堂里或实验室的显微镜旁，因此学生、研究者和老师之间会产生真正意义上的创造性互动。因此，总的来说大学校园的特色增加了这所大学的魅力。

于默奥大学的校园由 23 000 $m^2$ 的阳光甲板、堤岸、开放式草坪、步道和阶地组成。这些设施是围绕一个人工湖而组建的。湖中的一个小岛代表着通往一个小群岛的起点。这个群岛上有桥梁通向南部海岸。这里游客可以看到的是有很多小山丘的景观。山谷处有阳光也有阴影，散布于桦树的白色树干之下。

在活跃的学生活动中心前面，有一个向阳的户外休息区。这个休息区是一系列铺了碎石的扇形的阶地，每一处阶地都有咖啡馆装饰，并且有枝繁叶茂的树木提供遮蔽和阴凉。这个新的大学校园是 2007 年竞争的结果。

# Ningbo R&D Park (site B)

## 宁波研发园 B 区

Landscape Architect: Ohtori Consultants

The spatial arrangement is composed of two axes in the east-west and north-south directions, and four parks. The east-west direction is intended to form an open space axis of lawn, and the north-south direction to form a promenade axis of flowers and water.

Also, the four parks are given themes of "Swelling of blue waves", "Continuity of ripples", "Spreading of water rings" and "Still water".

Although the design motif of "water and waves" naturally necessitates varied types of waterscapes, the designers did not consider it sufficient to express the motif. It is intended that wherever people are in the premises, they are made to think of the water and waves by exploring two ways of approaches; namely, expressing each garden as a bold spatial composition and incorporating the water and waves into landscaping furniture, pavement design and other details.

Also, the design efforts were directed to making a difference in quality through spatial composition by incorporating different materials rather than pursuing only spatial variation, since the availability of construction materials in the current China might not be fully assured.

整个空间格局主要以东西轴和南北轴及 4 个公园组成。东西轴设计为大型草坪，营造敞开的景观空间，反之南北轴则利用花木及水景营造移步怡景之乐趣。

4 个公园均加于各自的特色，以“碧波荡漾”、“涟漪阵阵”、“水圈蔓延”、“净水深深”主题进行了设计。

整个场地的设计语言为“水与波”。为衬托主题在各个区域设置了不同形式的水景，但设计师们认为单靠实体形水景不足以表达他们的理念。因此将水的特色融入到景观设施、铺装及一些细节当中，使人们无论在哪个角落都能感受到水与波的设计理念，也为每个空间增加了独自个性。

由于中国现有的建筑材料可能无法完全满足设计需求，在设计过程中致力于搭配现有的材料来提升空间效果及质量，而不是一味的最求多样性。

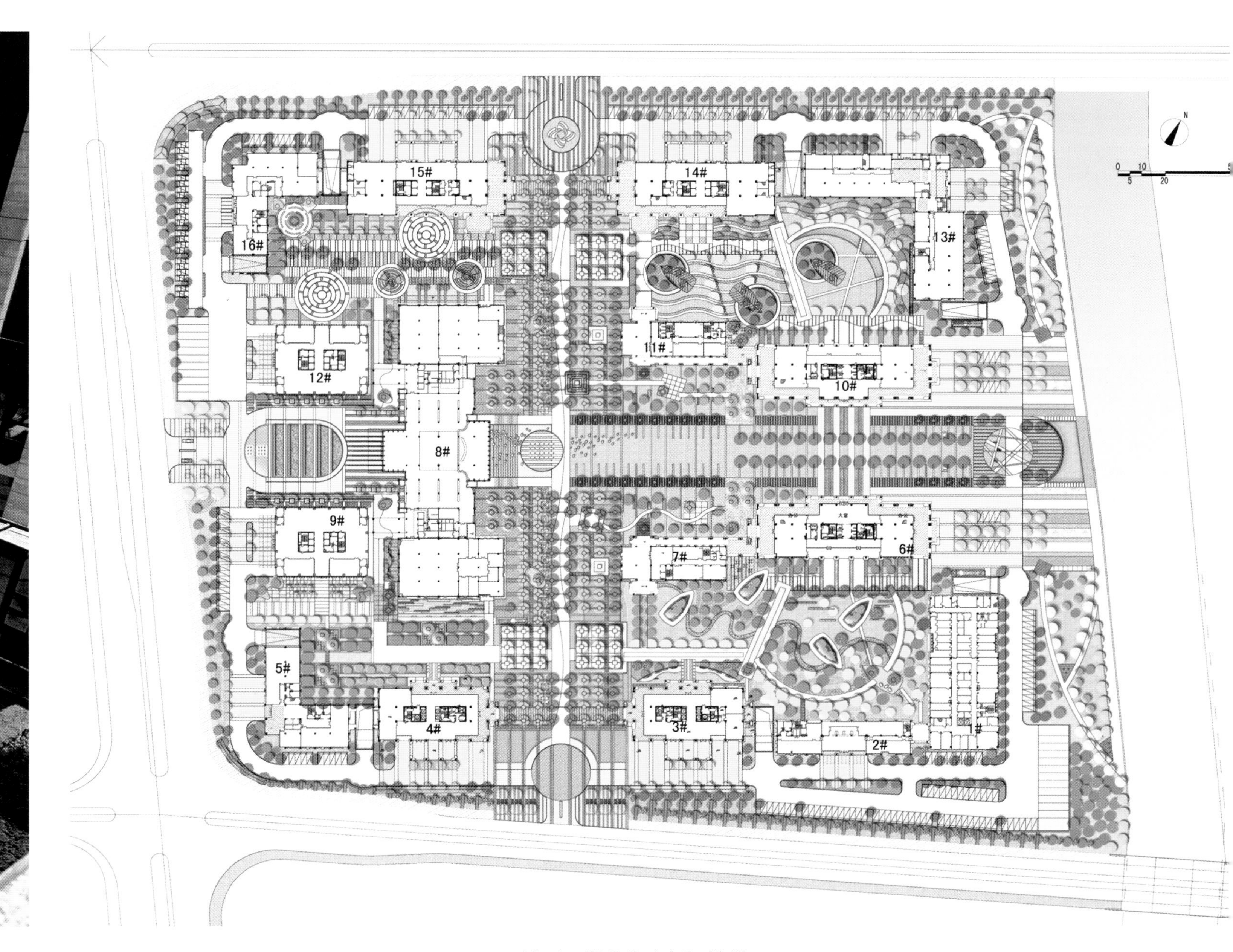


Ningbo R&D Park (site B) Plan

# Development Bank of the Meurthe

## 默尔特河岸开发区

Landscape Architect: Atelier CitE Architecture

The town of Raon l'Etape encloses a green area along the banks of the river Meurthe. This area, abandoned for a long period of time, is at the heart of a development project awarded by the town. The area of intervention defined by the project is delimited by the urban course of the river Meurthe. By developing both public and natural spaces the project will ensure a sustainable future for this zone.

The strategic position of this area allows the project to impact in several ways:
By unifying the different entities which make up the town of Raon l'Etape and thus involving the whole community;
By developing and installing new ways of transportation permitting a true alternative to the domination of the town by cars;
By providing a potential base for leisure activities and tourism;
Finally, the new development encourages a continuity and enhancement of the links between the town of the future and its cultural heritage.

Development Bank of ehe Meurthe Plan

在 Raon l'Etape 小镇里，默尔特河岸边有一块绿地。这块被遗弃很久的区域成为了小镇授予的开发计划的关键所在。这个项目的规划区域由默尔特河切分开来。这个项目通过开发公共空间和自然空间来保证环境的可持续发展。

该地区的战略地位使得这个项目有以下几个方面的影响：
将组成 Raon l'Etape 小镇的几个部分统一起来，带动整个地区的发展；
开发应用新的交通方式，替代小镇里占主导地位的汽车；
为休闲活动及旅游业提供潜在的场所；
这项开发最终要鼓励加强城镇未来与文化遗产之间的连续性联系。

这项开发因地制宜，且通过一些保守的影响突出了这个区域的独特性，彰显了这里的氛围。

第一个阶段保证该地区连续的城市中心形态，第二个阶段强调创造可提供休闲活动的自然景观，同时也要开发当地的新功能：
一个提供休闲活动的大型海滩和一个可以放松自我、防洪，也可以划独木舟的区域；
平缓的行人道带动了 Raon l'Etape 居民对河岸的使用。

The development fits in with the town's geography and uses discreet effects to highlight the uniqueness and the atmosphere of the area.

The phase 1 confirms the zone's continuity with city center's morphology while the phase 2 underlines the presence of leisure activities natural landscape but also proposes new use of the area:

A large beach for leisure activities and a space to relax also serves as a zone for flood control and for the creation of a canoeing area;

Gentle walkways encourage the use of the river banks by the inhabitants of Raon l'Etape.

# Sandgrund Park

## 桑德浅滩公园

Landscape Architect: Thorbjörn Andersson with Sweco architects

Design Team: Johan Krikström, Jimmy Norrman, Emma Pettersson, Lisa Hellberg

Client: City of Karlstad

Location: Karlstad, Sweden

Area: 40,000 $m^2$

This public park has a spectacular setting — a pointed peninsula in the shape of a birds beak, situated at a river bifurcation in the middle of the city of Karlstad, Sweden. The park is designed as a rolling landscape of long, parallel ridges, creating viewpoints at their tops and enclosed valleys in between. Each second valley is an activity field, and the others are plant habitats with a varied botanical content. The ridges are placed perpendicular to the peninsula to obscure the long view and instead reveal the attractions of this water landscape step by step. Along the river shores sun decks and jetties have been placed; spots for a rest during promenades or places from which one can enjoy the river.

Sandgrund Park Site Plan with Legend

这个公共公园的环境非常壮观，瑞典卡尔斯塔德中部河流分叉处有鸟喙形的尖头半岛。公园被设计成平行山脊般绵延起伏的景观，这样使得它们的顶部创造出新的视觉，而顶部之间形成了封闭的山谷。每隔一个山谷是活动场所，其余的山谷是栽种各种各样植物的地方。这些隆起线设置成与小岛垂直来模糊远景，一步步展现水体景观的魅力。河滨沿岸设有阳光甲板和码头，人们可以在散步时停留休息，或在这里欣赏河流美景。

# Le Grand Stade

## 跑马场景观

A key part of this relationship with the landscape is a pedestrian circuit leading around the whole of the site. This circuit is a public walkway, punctuated with elements of both the equestrian centre and the forest. Beginning at the entrance, the pathway crosses the exhibitors area, runs the length of the horses rest area, passes on a footbridge above riders on horseback, dips under the pine trees, runs alongside the collecting ring then the water jump, falls in line with the terraces, before climbing gently up to the top of the roof of the building, giving visitors plunging panoramic views over the arenas.

Staircases at regular intervals along the rooftop boardwalk give access to the stands and link to ground level circulation. This pedestrian circuit creates a loop around all of the Grand Parquet's paddocks and sand arenas, collecting them together in a sort of forest clearing dedicated to sport.

To the south, at the entrance, a sloping wall of climbing plants — a direct continuation of the grass slopes to either side — creates a visual and solar screen for the building behind.

The building's shape and the use of materials are also part of this desire to integrate the project into its surroundings.

Landscape Architect: Joly&Loiret

Client: Communauté de communes de Fontainebleau-Avon

Location: France

景观中关键的一部分是一条环绕整个区域的行人线路。这条线路是一个公共的散步道，穿过马场中心和森林。线路以入口处为起点，穿过展览中心，长度与马的休息区相当，经过竞技场上方的人行桥，途中种植了松树，线路沿着集电环前行，接着遇到水沟障碍，然后延伸到阶地方向，最后缓缓爬上建筑物的顶部，为游客提供一个急剧下降的竞技场全景的视野。

顶部木板路上的有一定间距的阶梯通向看台和地面上的环线区。这条行人通道创造出一个 Grand Parquet 围场与竞技沙场之间的循环线路，将它们集合到运动专用的森林空地中来。

南部入口处有一堵倾斜的爬满藤蔓的墙——每边草地斜坡的附加部分——为后面的建筑物营造出一片视觉上的防晒屏。

建筑物的外形和材料的使用旨在将项目融入到周围环境中去。

GENERALI

# Open-air Exhibition Grounds of the Estonian Road Museum

## 爱沙尼亚公路博物馆的露天展台

Landscape Architect: Salto Architects

Design Team: Maarja Kask, Karli Luik, Ralf Lõoke, Pelle-Sten Viiburg

Client: Estonian Road Museum

Location: Varbuse, Põlva county, Estonia

Area: 13,000 m$^2$

The concept of the additional outdoor exhibition area is based on a road – while strolling through, your route will be surrounded by different landscapes. The chosen solution forms a long 8-shaped path, where functions with different character and scale are placed in succession like a comic strip. The exhibition begins with an overview of traffic signs through history, continuing with segments of different types of historical roads, lined with objects related to travelling by road, as well as all kinds of machines used for maintaining or repairing them. A late-19th century steel bridge has been removed from its original location and given a new use linking the two parts of the exhibition above the entrance area.

Nearly all space necessary for the museum is scooped into the hilly South-Estonian landscape, leaving the rest of the environment as untouched as possible: natural and artificial landscapes are clearly separated. A hollow ranging from 10 cm to 4 m deep forms more than 13,000 $m^2$ of open-air exhibition space which is barely visible from the remote surrounding areas. For the most part, the structure is built of reinforced concrete, with wood-panelled 'nests' (ticket and souvenir booth, picnic area) and concrete walls with printed graphic images depicting roadside landscapes and scenes softening the object that is in itself a piece of infrastructure as well as architecture.

Phase two of the museum extension foresees the reconstruction of an existing exposition hall. Currently an inapt eyesore, the hall is to become fully integrated with the rest of the complex. In addition to exhibition space, it will also include a blackbox theatre and an office.

户外展览区这个概念是以公路为基础，当你漫步其中，你会发现沿途被各种不同的景观包围着。这个设计方案规划出一条 8 字形的小路，功能规模各不相同，如同一幅连环漫画。展览的起点是历史上的交通符号概况，接着是历史上不同类型的路段，排列着与公路旅行相关的物品，还有各种公路维修的机器。一条 19 世纪晚期的钢桥被迁移到这里，将入口区域的两个展览区连接起来。

几乎博物馆所有必需的空间都应用到了多山的爱沙尼亚南部景观中，尽量不破坏其他环境：自然景观与人工景观明显地分开。一个 10cm 到 4m 深的山谷区形成了 13 000$m^2$ 的露天展览区，只有从远处才看得见。这个建筑结构从很大程度上来说是由钢筋混凝土构成，且设有一些木板“巢”（出售入场券和纪念品的摊位、用餐区）和一些印有图案的混凝土墙壁，描绘着路边的风景，柔和了那些本身既是基础设施又是建筑的物体。

博物馆扩建的第二阶段规划了一个现存的展厅的重建项目。目前这个展厅与整个景观显得不协调，重建后它将会融入整体。除了展览空间，这里还会建一个小剧场和一个办事处。

# CONTRIBUTOR / 设计师名录

## AECOM

AECOM's Design + Planning practice fuses landscape architecture, masterplanning, and environmental planning to facilitate responsible and productive use of land. Working from studios across the world, they have designed public realm in Tokyo, London, New York, Los Angeles, and dozens of places in between. Their designs create engaging, healthy, and safe environments for people while protecting and restoring natural systems and encouraging the sustainable use of natural resources. Often associated with waterfronts, transportation infrastructure, or central business and shopping districts, these landscapes play a key role in the life of their cities and citizens.

## Andropogon Associates Ltd.

Founded more than thirty years ago, Andropogon is committed to the principle of "designing with nature," creating beautiful and evocative landscapes inspired by the careful observation of natural processes and informed by the best environmental science. The elegance and economy of natural form and process continues to be the benchmark by which we measure the success of our work – from the smallest construction detail to the multi-layered patterns of regional sites.

As a certified minority business enterprise (MBE), Andropogon is committed to diversity and inclusiveness in the workplace. Their multi-cultural staff is dedicated to the successful maturing of each project, from initial concept designs to construction review and long-term landscape management. Their body of national and international work includes early examples of innovative green strategies that have withstood the test of time as well as a broad range of landscape, site planning, environmental projects, ecological restoration and innovative stormwater management techniques.

Andropogon's clients often tell them that they combine integrated design with a depth of ecological understanding in synergistic ways. With every project they embody their mission... "to weave together the landscapes of man and nature for the benefit of both".

## ASPECT Studios

ASPECT Studios is a group of design studios united through a philosophy that delivers innovative landscape architecture, urban design and digital technologies.

Since its beginning, ASPECT Studios has grown to over 80 people throughout their Australian and China studios. They have established a reputation for design-led solutions and are recognized as a company with the capability to deliver world leading design excellence through creative and sustainable projects.

## Atelier CitE Architecture

Atelier CitE Architecture is a workshop whose field of investigation crosses city territory and architecture. In very diverse professional practices (diagnosis, project teaching, research and publication), the different partners of city architecture have always had the desire to develop global approaches and multidisciplinary territories and give great importance to the development of diagnostic to highlight the elements of identity of the place and build a contemporary project.

Thinking about the long-term impacts of decisions on development, the environment and landscapes have always been the backdrop of all their work: the question of urban sprawl, the effects of "the all for car" the treatment of urban networks (travel, water, energy, infrastructure) and the identity of each place.

Dominique Cico: National Architecture Diploma at Architectural school of Paris la Villete, Environment and Landscape Diploma at the Architectural School of "Paris-la-Villette", Paris — France.

Bruno Tonfoni: National Architecture Diploma at Architecture School of Grenoble, France; Urbanism Diploma at the National School of "Ponts et Chaussés" (ENPC); French State Adviser Architect; Professor of architecture and urban design at the Architectural School of "Paris Val de Seine" at Paris, France.

## Ramboll Studio Dreiseitl

Ramboll Studio Dreiseitl is an award-winning, sustainable landscape architecture practice. They are a multidisciplinary team specialized in the synthesis of landscape architecture, art & urban design, environmental technology and urban hydrology. With a total of four studios located in Germany (Überlingen and Hamburg), Singapore and China (Beijing), their 100 multicultural experts are landscape architects, urban planners, engineers, architects, and artists working on the daily reality of transforming our cities into resilient homes.

35 years of experience has created a portfolio of outstanding projects with high aesthetic and cultural value, claiming several major international and local awards such as the World Architecture Festival Award, American Society of Landscape Architecture Award, Singapore Landscape Architecture Award and President Design Award. They are a member of the Singapore Institute of Landscape Architects, and their scope of the practice's work includes water sensitive urban design, streetscapes, masterplans, parks and plazas, and down to the applied scale of swales, biotopes and building-integrated rainwater recycling systems.

## Carve

Carve is a design and engineering bureau that focuses on the planning and development of public space, particularly for use by children and young people. The bureau was set up in 1997 by Elger Blitz and Mark van der Eng. Carve considers playgrounds an integral part of public space. They strive for playgrounds to be inviting places to be discovered; providing room for different groups and ages, and several forms of use. They want their playgrounds to be challenging and safe at the same time. Carve's leading motive is to create space for play and to foster possibilities. Over the past seventeen years Carve has grown into a company within which several design disciplines meet, from industrial design to landscape architecture.

## DS Architecture – Landscape

Being a pioneer in its field since 20 years, DS is a design office that has both theoretical and practical products on where architecture and landscape intersect. The main goal of the works is to create sustainable, practicable, well designed places with something to tell.

DS designs have the "setting of a micro cosmos" as the main motivation. The design group perceives the complex fabric of today's multilayered, multicultural and interdisciplinary environment as a source and using this source, it sets up meaningful, clear, creative new places with the motto of combining high artistically skills with rational, economic and ethical values; both in national and international projects.

Considering nature as another source, the group creates parametrical designs thorough a sentactical perception of nature by filtering it.

DS provides design services ranging in scale from residential and urban projects, cooperating with distinguished architects, planners, engineers and especially ecologists.

## glasser and dagenbach, landscape architects bdla, IFLA

Silvia Glasser and Udo Dagenbach founded their partnership in 1988. Since then they created innovative and highest quality parks and landscapes, but also especially public places and private gardens. Silvia Glasser is a state approved gardener specializing in perennials and received her Diploma in Landscape Architecture by the University of Nuertingen in 1985. Udo Dagenbach is a state approved gardener too with a Diploma in Landscape Architecture from the Technical University of Berlin in 1986. He also studied stone sculpture at the University of Art in Berlin as a guest student of professor Makoto Fujiwara and worked with him. The small office with three landscape architects and two to three assistants primarily focuses on the new design of public parks, Hotels and Resorts and private residential projects as well as the reconstruction of gardens and parks. Their philosophy: The settings of gardens and parks form backdrops before which visitors, whether public or private, are able to act out a role in their very own play.

The projects are spread over Europe , Russia, Georgia, Armenia and Azerbaijan.

The office received several awards such as Deutscher Landschaftsarchitektur-Preis 2007 of the BDLA (1st prize), daylight spaces award 2007 / international architecture and design competition (1st prize) and Made in Germany — Best of Contemporary architecture (2nd prize).

## Grant Associates

Grant Associates is a world-leading British Landscape Architecture consultancy specialising in creative, visionary design of both urban and rural environments worldwide, working with some of the world's leading architects and designers.

Inspired by the connection between people and nature Grant Associates fuses nature and technology in imaginative ways to create cutting edge design built around a concern for the social and environmental quality of life.

Grant Associates has experience in all scales and types of ecological and landscape development including strategic landscape planning, master planning, urban design and regeneration and landscapes for housing, education, sport, recreation, visitor attractions and commerce.

## Groupe IBI-CHBA

Formerly known as Groupe Cardinal Hardy, Groupe IBI-CHBA is most actively engaged in the rebirth of the Montreal, Canada. The team led by seven associates has acquired a solid reputation, placing it among the most important professional practices of its kind in Canada. The firm specializes in landscape architecture, urban studies, architecture, restoration and recycling of historical sites and buildings, transportation, mixed-use projects and housing. Groupe IBI-CHBA is renowned for its rigorous studies in urban planning, for its multidisciplinary approach and its over 70 awards of excellence.

## In Situ

The In Situ agency brings together a team of fifteen (landscapers, architects and planners) around Emmanuel Jalbert, landscape and urban planner. In addition to his project management activity, Emmanuel Jalbert works as a consultant and is also involved in teaching and research.

For over 20 years, In Situ has led multiple development projects. It has particular expertise of public space in all its forms: parks, squares, gardens, plazas, docks and riverbanks. Their work extends from dense city to wild landscape and each project aims to combine nature and culture, to reconcile memory and modern times.

## Joly&Loiret

Joly&Loiret was founded by Paul-Emmanuel Loiret and Serge Joly in 2006. Since then, the office has had between 6 and 8 employees. Work has diversified both in the public and private sector, notably for schools, sporting and cultural public commissions, but also in exhibition design, housing and offices

In the manner of a landscape designer or land artist, the office's projects are more than simply functional. They are cultural, determined by a sense of place, its history, its form, its materials and its textures with the aim of creating a space to fit the local context. Each project, conceived with respect for the use of energy and resources, reveals and reflects its environs, as if it had risen from nature.

Most importantly, the office's projects aspire to be sensitive, efficient, dynamic, sustainable, appropriate and alive, to be poetic, to touch the people who use them. Far from having a formal ideology, Joly&Loiret strive to build an architecture that is concerned not with its own singular presence, but with the actions and events that it might engender.

## LANDLAB studio voor landschapsarchitectuur (Arnhem)

LANDLAB studio for landscape architecture is a landscape architecture firm that creates functioning and experientially rich places across a wide range of landscape scales: from landscape to park to garden. LANDLAB has extensive experience in the design of streets and plazas, parks, zoo's, roof gardens and landscapes. Several projects such as Funenpark, the plaza in front of the EYE film institute and the Oeverpark are critically celebrated for their achievements. LANDLAB won prizes in Belgium for two shopping streets, De Driekoningenstraat and De Abijstraat, and in received an Dutch award for their work on the Funenpark.

## LOOK Architects Pte Ltd.

Founded in 1993, LOOK Architects is a design-intensive practice committed to rigorous analysis and research to produce innovative, iconic buildings and urban design.

Founders Look Boon Gee and Ng Sor Hiang foster a collaborative studio environment to germinate and develop design ideas on a broad spectrum of works in the Southeast Asian region, defining LOOK Architects as a practice keen on breaking new ground in various fields of design. Recipient of The President's Design Award – Designer of the Year 2009, Look Boon Gee's passion for his craft continues to lead the practice towards exciting directions.

Versatility in handling projects of vastly different natures is evident in completed works, an array that includes private residences, apartment tower, university institution, community library, commercial headquarters, waterfront promenade and bridges. Embracing a design philosophy of the integrated whole, a high degree of inventiveness goes into construction detailing, ecological use of building material and planning of spatial relationships in relation to the environment.

Clients' personal and collective experiences are creatively interpreted to form perspectives that enrich the design approach of each project, and together with sustainable responses to the distinctive character of place, design enters the realm of eco-poetry, a lively dialogue that reverberates between nature and aspirations of man.

## L&A Design Group

L&A Design Group is one of the leading and largest companies dedicated to landscape design, urban planning and architectural design in China. L&A was established in 1999 by Mr. Bo Li, a Canadian registered landscape architect, and moved to Shenzhen in the year of 2001. Today, L&A has offices in Shenzhen, Shanghai, Beijing and Xi'an, together they serve clients all over the great regions of China. L&A follows the design principle of landscape urbanism and believes in an integrated approach in the solution of China's complicated urban issues.

Till now L&A has accomplished over 600 reputable landscape, architectural and urban design projects in China with many won national awards. L&A takes pride in its dynamic international design team, with a multi-disciplinary team of more than four hundred professionals including registered architects, landscape architects, planners, engineers, economists and artists.

## Mayslits Kassif Architects

Mayslits Kassif Architects was founded in Tel Aviv in 1994 by Ganit Mayslits Kassif and Udi Kassif. Since 1994 the practice is involved in a variety of projects in the fields of urban planning, public buildings, housing, retail and landscape urbanism, all reflecting the firm's view of architecture as an agent of urban and environmental transformation.

Since its inception the practice have won several major public competitions and constructed some significant architectural projects such as the renewal of Tel Aviv Port, the renovation of the Central Promenade of Tel Aviv, the renewal of Acre waterfront and the Remez Arlozerov community center. The practice was awarded several prestigious architectural awards such as The Rosa Barba European Landscape Prize in 2010, as well as the Rokach Award and the Rechter Award, for remarkable architectural projects in Tel Aviv.

## Meyer + Silberberg Landscape Architects

Meyer + Silberberg's collaborative design studio is based in Berkeley, California. They are a team of eight dedicated professionals passionate about the design of the landscape. David Meyer and Ramsey Silberberg have taken great care to shape their practice into a uniquely responsive and personable enterprise. Coming from some of the most admired practices in the country, they bring over 40 years of knowledge and experience to every project. The firm services a wide range of clients, including non-profit institutions, private developers, academic campuses, business associations, and government agencies. What unites the work is the premise that there is always something inherent in a site and the surrounding culture that wants to be expressed. So they express it with distinction and with simplicity. They craft landscapes that transcend and anchor themselves in the hearts and heads of the people who use them.

Meyer + Silberberg is recognized internationally for the increasingly rare ability to transform a great idea into an exceptional physical space. The firm does this through passionate engagement with their clients, tireless exploration and refinement of design, and a renowned reputation for construction and execution.

## Mikyoung Kim

Mikyoung Kim is an award winning international landscape architect and artist whose work focuses on merging sculptural vision with the urban landscape. As principal and design director, Mikyoung Kim has brought her background in sculpture and music, as well as her design vision as a landscape architect, to the firm's diverse work. Projects are comprised of designs that meld site, sculpture and sustainable initiatives to develop engaging and poetic landscapes. Over the past five years, Mikyoung has been involved in projects at various scales that focus on the choreographed experience with the use of a wide range of technologies with light and color. Since the firm's inception, the work of Mikyoung Kim has received critical acclaim winning multiple national awards from the American Society of Landscape Architects, the American Institute of Architects, and the International Federation of Landscape Architects, as well as awards from the Boston Society of Architects and the Boston Society of Landscape Architects. Recent work has been featured in numerous publications, including Architectural Record, the New York Times, the Washington Post, Sculpture, Dwell, Surface Magazine Landscape Architecture, Land Forum, Garden Design, Interior Design, Pages Paysages, and a monograph of the firm's work in Inhabiting Circumference. Mikyoung is also Professor of Landscape Architecture at the Rhode Island School of Design.

## Nyréns Architects

Nyréns is an award winning Swedish architecture practice where the majority of the employees are shareholders of the company. The office was founded in 1948 by Carl Nyrén and today we are around 120 people working in two offices in Sweden: in Stockholm and Malmö. Nyréns employs architects, engineers, interior designers, landscape architects, city planners, architectural conservationists, 3D visualisers and architectural modelers. We look for integrated solutions that bring urban design, master planning, building design, engineering, landscaping, interior design and architectural conservation expertise together.

Democratic co-ownership, small design teams dedicated to each project and a flat organisation make Nyréns a place of equality and enthusiasm. All teams have a cross-section of disciplines, including architects, interior designers, engineers, landscape architects, master planners and building antiquarians, with co-ordination between the groups providing an invaluable breadth of skill and experience. Given Nyrén's philosophy to involve the clients in the architectural process, direct contact between our clients and a key contact within the team is established for every project to achieve close collaboration and ease of communication.

## Ohtori Consultants Environmental Design Institute

This institute consists of a group of designers specializing in landscape architecture. They deal with all kinds of environmental design including landscape planning and architectural design, and always strive to propose and realize the creation of new cities and regional environment incorporating a variety of fields from engineering, landscape gardening, and ecology into our design fields. The content of their business ranges from urban development, commercial space, plaza, living environment, green land of park to garden design. In recent years, they have been engaged in many overseas projects as well.

## Rainer Schmidt landscape architects + city planners

Rainer Schmidt landscape architects + city planners have been designing and implementing projects on different scales for 25 years in Germany and abroad, specializing in the regions of North Africa, Middle East and Asia. The firm employs a staff of approximately 30 and has three offices in Germany: the head office is in Munich; one branch office is in Berlin and the other in Bernburg. The office is specialized in the planning and implementing of largescale projects in landscape architecture, environmental planning and urban design. The office's aim is to find suitable approaches to solve the problems of our time. The language of landscape architecture in the 21st century must offer a realistic reflection of the way people interact with each other and with nature. The office is striving to achieve a balance between design, function, emotion and conservation.

Rainer Schmidt is Professor for Landscape Architecture at Technical University of Applied Sciences Berlin since twelve years, and has been guest professor at Beijing University in 2005 and Farrand Professor at University of California, Berkeley in 2007. Since 2005 Rainer Schmidt is vice president of the German society of garden architecture and landscape culture (DGGL).

## Ravetllat Ribas arquitectes

Pere Joan Ravetllat and Carme Ribas are architects from Barcelona School of Architecture (ETSAB) since 1980, and started developing their professional career together in 1985. They focused their work in urban transformation projects, residential and public space. Their projects always try to give an urban perspective: housing, educational facilities, refurbishment of old buildings or landscape projects are always worked in relationship with the city.

From the beginning they combined its professional activity with teaching and researching in ETSAB. Indeed, from their point of view, these two activities are complementary and helped them to achieve different points of view on the profession.

## Salto Architects

Maarja Kask (b 1979), Ralf Lõoke (b 1978) and Karli Luik (b 1977) founded Salto in Tallinn, Estonia in 2004. It is an office devoted to architectural practices from interior design and art projects to landscape architecture and urban planning projects. In recent years Salto has gained prizes in over 40 architectural competitions in Estonia and abroad.

## scape Landschaftsarchitekten

The office scape landscape architects was founded in 2001 by Matthias Funk, Hiltrud M. Lintel and Rainer Sachse in Duesseldorf. For our clients we are working with a young and motivated team, in cooperation with city planners, architects, ecologists and communication designers mainly on object-planning designs of urban open spaces. The current projects range from master development plans for entire districts over conceptual designs for parks, pedestrian streets and squares, through to detailed planning of our own street furniture systems.

## SLA

SLA is an internationally renowned landscape architecture firm that works with landscape, urban space and city planning. SLA creates modern, sustainable urban spaces that inspire community and diversity through innovative use of architecture, infrastructure, nature, design and technology.

SLA developes urban spaces in all scales, from masterplans to installations, in Denmark and elsewhere. SLA explores innovative methods of design and research, and has developed distinctive approaches on themes such as participatory planning and innovative preservation.

The office was founded by architect Stig L. Andersson in 1994 and has since won numerous competitions on landscape and urban design in Denmark and abroad. Stig L. Andersson is owner and Creative Director, Lene Dammand Lund is the Managing Director and Hanne Bruun Møller Associate Partner.

SLA is a member of Danish Association of Architectural firms (Danske Ark) and our landscape consultancy services are provided in accordance with the principles stated in their rules and regulations.

## Surfacedesign, Inc.

Surfacedesign, Inc. was established in 2001 to provide clients with a broad range of professional design services, including landscape architecture and urban design and master planning. The award-winning practice is engaged in projects of a variety of different scales, both locally and internationally: estate design, park design, hospitality, corporate campuses, municipal streetscapes, and large-scale land use planning and urban design projects.

Surfacedesign create projects that have a strong relationship to people and the natural environment, they are passionate about craftsmanship and sustainability.

## SWA Group

For over five decades, SWA Group has been recognized as a world design leader in landscape architecture, planning and urban design. Their projects have received over 600 awards and have been showcased in over 60 countries. Their principals are among the industry's most talented and experienced designers and planners. Emerging in 1959 as the West Coast office of Sasaki, Walker and Associates, the firm first assumed the SWA Group name in 1975.

Despite being one of the largest firms of its type in the world, SWA is organized into smaller studio-based offices that enhance creativity and client responsiveness. Over 75% of their work has historically come from repeat clients. In addition to bringing strong aesthetic, functional, and social design ideas to their projects, they're also committed to integrating principles of environmental sustainability. At the core of their work is a passion for imaginative, solution-oriented design that adds value to land, buildings, cities, regions, and to people's lives.

SWA provides complete landscape architectural services, including site planning, concept design, schematic design, design development, construction documentation and construction observation. They often provide our clients with continuing landscape design consultation after construction completion and can provide landscape management plans for their use.

SWA producse comprehensive land plans and master plans for large land areas. Because their work is land-based, they're able to create plans that sensitively make the best use of terrain, landform, natural systems, landscape, and urban spaces, and integrate those elements with the required infrastructure, buildings, and other improvements. They apply these same skills to projects involving the use and restoration of natural systems.

SWA offer master planning, preparation of design guidelines, and full design services for urban projects. These urban design and planning services can be applied to entire districts, as well as street systems, city blocks, public parks and plaza spaces, waterfronts, and the smallest of urban areas. They are skilled at addressing both the redevelopment of an urban area—including infill development and land use changes—as well as the design of new urban environments.

## Thorbjörn Andersson with Sweco architects

In the field of landscape architecture, the studio concentrates on public space in urban environs – basically parks and plazas. The philosophy is that quality of civilization can be measured by how well society can handle what we have in common, and public space is certainly something that all of us share. There is no other place where this becomes more evident than in the public realm. Parks, plazas and streets is where urban life is. Public space is the training arena for tolerance in the multi-cultural society. Urban culture needs public space to show us our shortcomings as well as our assets. The Hyllie Square in Malmö is an evident example of an urban square that collects visitors from all corners of society. The Campus Park at Umeå University is in some ways the opposite: here, only people active at the university meet. Still, the park is the most important space of that university; probably more important than the lecture halls or the separate laboratories. It is in the open space that encounters happen, people meet and ideas are being born.

Sweco architects is a publicly owned firm with about 500 employees distributed over about 30 different offices in northern Europe. The landscape studio in located in Stockholm and from this base the studio operates and engages itself in projects all over the world.

## Tom Leader Studio

Tom Leader Studio (TLS) is an award winning landscape architecture practice with offices in California and Minnesota. Since its inception in 2001, the studio has invested and built upon the unique, inherent qualities of cities and their landscapes. In so doing, TLS seeks to provide a link between e merging ideas and practices and the concrete need for their realiza tion in physical space.

## Townshend Landscape Architects

Townshend Landscape Architects is an award winning practice formed in 1988 comprising an international team of landscape architects, urban designers, architects and artists.

Their approach to design is embedded in the local distinctiveness and character of each and every scheme and encompasses an essential view on sustainability across all the environmental, social and economic dimensions. With a deep interest in the history and formation of public spaces they have developed an understanding of how thriving and sustainable communities evolve. This commitment to successful place-making is at the core of the rigorous design process, connecting to the often fascinating history of a site, the surrounding built environment and to the people that will use them. It is these factors that inform their approach to creating unique, enduring places which respond to the inherent qualities of a site and the communities surrounding them, in turn creating opportunities for further investment.

The practice specialises in designing landscape, urban design and the public realm. They cover all aspects of a project, from inception, local authority and public consultations, planning strategies and gaining planning permissions, concept and detailed design, construction and the successful delivery of a project.

The individual and well thought through designs, coupled with an extensive knowledge of successful enduring planting palettes show passion, commitment and a deep understanding of how to create beautifully crafted and executed places for people around the world.

## Wachman Architects

Wachman Architects is located in Ein-Hod Artist's Village on Mount Carmel. Since its foundation in 1995 (first as Wachman-Brunovsky), its comprehensive architectural practice includes housing and interior design, public buildings, urban design and landscape architecture.

In recent years, alongside with other diverse projects, Wachman Architects are involved in the revival environment of the cooperative rural communities in Israel – the "kibbutz" and its surroundings.

In the last decade, parallel to running the architectural practice, Amos Wachman was teaching at the Faculty of Architecture and Town Planning, Technion, Haifa.

ARTPOWER

## Acknowledgements

We would like to thank all the designers and companies who made significant contributions to the compilation of this book. Without them, this project would not have been possible. We would also like to thank many others whose names did not appear on the credits, but made specific input and support for the project from beginning to end.

## Future Editions

If you would like to contribute to the next edition of Artpower, please email us your details to: contact@artpower.com.cn